▶ Concepts and Causes in the Philosophy of Disease

DOI: 10.1057/9781137552921.0001

Other Palgrave Pivot titles

Ellen McCracken: **Paratexts and Performance in the Novels of Junot Díaz and Sandra Cisneros**

Chong Hyun Christie Byun: **The Economics of the Popular Music Industry**

Christina Papagiannouli: **Political Cyberformance: The Etheatre Project**

Lorann Downer: **Political Branding Strategies: Campaigning and Governing in Australian Politics**

Daniel Aronoff: **A Theory of Accumulation and Secular Stagnation: A Malthusian Approach to Understanding a Contemporary Malaise**

John Mohan and Beth Breeze: **The Logic of Charity: Great Expectations in Hard Times**

Carrie Dunn: **Football and the Women's World Cup: Organisation, Media and Fandom**

David R. Castillo, David Schmid, Dave Reilly and John Edgar Browning (editors): **Zombie Talk: Culture, History, Politics**

G. Douglas Atkins: **Strategy and Purpose in T.S. Eliot's Major Poems: Language, Hermeneutics, and Ancient Truth in "New Verse"**

Christophe Assens and Aline Courie Lemeur: **Networks Governance, Partnership Management and Coalitions Federation**

Katia Pilati: **Migrants' Political Participation in Exclusionary Contexts: From Subcultures to Radicalization**

Yvette Taylor: **Making Space for Queer-Identifying Religious Youth**

Andrew Smith: **Racism and Everyday Life: Social Theory, History and 'Race'**

Othon Anastasakis, David Madden, and Elizabeth Roberts: **Balkan Legacies of the Great War: The Past is Never Dead**

Garold Murray and Naomi Fujishima: **Social Spaces for Language Learning: Stories from the L-café**

Sarah Kember: **iMedia: The Gendering of Objects, Environments and Smart Materials**

Kevin Blackburn: **War, Sport and the Anzac Tradition**

Jackie Dickenson: **Australian Women in Advertising in the Twentieth Century**

Russell Blackford: **The Mystery of Moral Authority**

Harold D. Clarke, Peter Kellner, Marianne Stewart, Joe Twyman and Paul Whiteley: **Austerity and Political Choice in Britain**

palgrave▸pivot

Concepts and Causes in the Philosophy of Disease

Benjamin Smart

Senior Lecturer, University of Johannesburg, South Africa

DOI: 10.1057/9781137552921.0001

First published 2016 by
PALGRAVE MACMILLAN

Palgrave Macmillan in the UK is an imprint of Macmillan Publishers Limited, registered in England, company number 785998, of Houndmills, Basingstoke, Hampshire RG21 6XS.

Palgrave Macmillan in the US is a division of St Martin's Press LLC, 175 Fifth Avenue, New York, NY 10010.

Palgrave Macmillan is the global academic imprint of the above companies and has companies and representatives throughout the world.

Palgrave® and Macmillan® are registered trademarks in the United States, the United Kingdom, Europe and other countries.

ISBN: 978-1-137-55293-8 EPUB
ISBN: 978-1-137-55292-1 PDF
ISBN: 978-1-137-55291-4 Hardback

A catalogue record for this book is available from the British Library.

A catalog record for this book is available from the Library of Congress.

www.palgrave.com/pivot

DOI: 10.1057/9781137552921

For MJT

DOI: 10.1057/9781137552921.0001

Contents

DOI: 10.1057/9781137552921.0001

List of Figures

▶

Acknowledgements

I am grateful to Kate Graham, Alex Broadbent, Pendaran Roberts, and especially to Michael Talibard.

Section 3 of Chapter 3 is reproduced with kind permission from Springer Science+Business Media. Smart, B.T.H. On the Classification of Diseases. *Theoretical Medicine and Bioethics* (August 2014), 35(4): pp. 251–69.

Figures 1.2 and 2.1 are reproduced with kind permission from The University of Chicago Press. Schwartz, P. H. Defining Dysfunction: Natural Selection, Design, and Drawing a Line. *Philosophy of Science* (July 2007): p. 272 and p. 378 © The Philosophy of Science Association.

DOI: 10.1057/9781137552921.0003

List of Abbreviations

BMI	body mass index
BST	biostatistical theory
CCD	causal classification of disease
CVD	cardio vascular disease
EBM	evidence based medicine
EBP	evidence based practice
ETPC	etiological theory of pathological condition
FNC	frequency and negative consequences (approach)
GP	general practitioner
NHST	null hypothesis significance test
POA	potential outcomes approach
RCT	randomised control trial

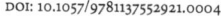

DOI: 10.1057/9781137552921.0004

palgrave▸**pivot**

www.palgrave.com/pivot

Introduction

Smart, Benjamin. *Concepts and Causes in the Philosophy of Disease*. Basingstoke: Palgrave Macmillan, 2016. DOI: 10.1057/9781137552921.0005.

▶

Disease is everywhere. Everyone experiences disease, everyone knows somebody who is, or has been diseased, and disease-related stories hit the headlines regularly. Just this morning, in fact, I heard that being tall increases one's risk of cancer. I am not a short man, so this did not please me. On the other hand, 30 seconds later, the same news reader announced that the key to curing cancer might just be found in the sting of an unusual species of wasp – so I left the house relatively content.

Unfortunately, when diseased, one often suffers pain and nausea, spends more time than usual in the bathroom(!), and is generally unable to do the things one enjoys. Eventually, of course, it is disease[1] that relieves us of our mortal coil.

Fortunately, if one is diseased, one can generally do something about it; namely, one can either go about treating oneself (perhaps by taking a pain killer), or when more serious, one can visit a clinician. If one is lucky, the clinician will be able to explain what is causing the symptoms, diagnose a condition, and prescribe treatment. Regularly, however, a single visit to your general practitioner (GP) is not enough for a diagnosis, since she will often have to send blood samples/urine samples/skin samples and so on for analysis, first. Here, the clinician asks for the pathologist's help.

Unlike the GP, the pathologist knows nothing of how the patient is feeling, or what the patient's future intentions are. Those qualities are relevant to the clinician, since she must decide how to proceed with treatment, but not to the pathologist. Very crudely, the pathologist's job is to identify the disease a patient is suffering from, by carefully testing/examining the physical evidence at her disposal (and on occasion, to determine what caused a patient's death).

Once the clinician can make a diagnosis (based on the pathologist's conclusions), she can prescribe treatment – but she would be unable to do this, without the vast quantity of medical research that occurs 'behind the scenes'; the clinician must know what the appropriate treatment *is*. Establishing the determinants of, and treatments for disease, involves designing and conducting all kinds of studies and trials, and doing lots of statistical analysis. This is the task of epidemiologist. Crudely stated, epidemiology is 'the study of health and disease in populations' (Saracci, 2010, 2), for the purposes of improving public health. Without it, clinical practice (at least as we know it today) could not get off the ground.

Unsurprisingly, much of the philosophy of disease literature targets a conceptual analysis of disease. However, there is more to the philosophy

DOI: 10.1057/9781137552921.0005

of disease than answering the question: 'what is disease?' Clinical medicine, pathology, and epidemiology, all carry with them a host of interesting philosophical issues. The target of this book is to provide the reader with an insight into some of the existing arguments in the philosophy of disease, and to build on these debates where I can.

In Part I of the book I follow the trend and consider what we mean when we talk of 'disease'. This question cannot, however, be answered from an 'in general' perspective, since as we have just seen, medicine comprises several, very different disciplines. I therefore analyse the concept of disease, first from the perspective of the clinician, and then from the perspective of the pathologist. In this short book there is not space to consider *every* argument in the literature. However, by focusing on the most prominent figures in this debate, I hope to provide the reader with a good understanding of the more influential views and their problems. I also propose two novel conceptions of disease, one for the clinician, and one for the pathologist. We shall see that the concept of the clinician and that of the pathologist, do not coincide – the answer to the question 'what is disease?' is thus entirely context dependent.

In Part II of the book I discuss causation in medicine. Since Hume, numerous analyses of causation have been offered by analytic philosophers. I apply several of these to medicine, discovering that those working in the medical professions use different conceptions of causation in different circumstances. Interestingly, we shall see that some conceptual analyses of causation are useless for clinical practice, but nonetheless crucial for improving public health. To end Chapter 3, I propose a model for classifying diseases by their causes.

The final chapter of this book concerns causal inference in public health. In this chapter, I analyse two theoretical notions employed by epidemiologists: the potential outcomes approach to effect measurement, and Kenneth Rothman's sufficient-cause model to epidemiology. The philosophy of epidemiology is (at the time of writing) a fledgling discipline. I hope this discussion provokes more research into a hugely underexplored area of philosophy.

Note

1   In the 'pathological condition' sense.

DOI: 10.1057/9781137552921.0005

Part I
The Concept of Disease

▶

DOI: 10.1057/9781137552921.0006

1

The Concept of Disease in Clinical Medicine

Abstract: *This chapter asks the question: 'what, for the clinician, is disease?' Of all the tasks philosophers undertake, I consider the conceptual analysis of disease to be of particular importance – and for obvious reasons. For one, one's disease status can have significant consequences. In the following sections I outline three prominent conceptions of disease, looking for that best suited to clinical medicine: those proposed by Rachel Cooper (2002), Peter Schwartz (2007), and Jerome Wakefield (1992, 1999). I dismiss Cooper's view based on clear cut counterexamples, but show that significant parts of both Wakefield and Schwartz's views are compelling. I conclude with a fourth proposal (which draws on both Wakefield and Schwartz), providing an etiological theory of disease in clinical medicine.*

Keywords: disease as harmful function; harmful dysfunction; Jerome Wakefield; natural function; Peter Schwartz; the line-drawing problem; What is disease?

Smart, Benjamin. *Concepts and Causes in the Philosophy of Disease*. Basingstoke: Palgrave Macmillan, 2016. DOI: 10.1057/9781137552921.0007.

Introduction

> [H]ow we think of health and disease lies at the very core of medical practice, reflections on bioethics, and the formation of health care policy.
>
> – Engelhardt in Ananth 2008.

This chapter asks the question: 'what, for the clinician, is disease?' Of all the tasks philosophers undertake, I consider the conceptual analysis of disease to be of particular importance – and for obvious reasons. To mention just a few: concepts of health, illness, and disease (specifically mental illness) affect whether or not one is legally responsible for one's actions; a concept of disease is necessary to demarcate the diseased from the healthy (which might distinguish those eligible for medical treatment, from those not[1]); not only must medical care be paid for, but one's state of health affects one's right to disability allowance, free housing, and so on – so concepts of health and disease have a significant economic effect; and since the decisions made by those in the medical profession affect legal responsibility, eligibility for medical care, eligibility for income support, and the distribution of resources (both locally and internationally), decisions concerning health and disease have enormous ethical implications.

In the following sections I outline three prominent conceptions of disease, looking for that best suited to clinical medicine: those proposed by Rachel Cooper (2002), Peter Schwartz (2007), and Jerome Wakefield (1992, 1999). I dismiss Cooper's view on the basis of clear cut counterexamples, but show that significant parts of both Wakefield and Schwartz's views are compelling. I conclude with a fourth proposal (which draws on both Wakefield and Schwartz), providing an etiological theory of disease in clinical medicine, that is not, unlike Wakefield's etiological account, subject to the line-drawing problem.

The maximally value-laden conception – Rachel Cooper on disease

Cooper provides a tripartite account of disease; that is, she proposes that the concept of disease has three individually necessary, and together sufficient conditions.

DOI: 10.1057/9781137552921.0007

1   Disease is a bad thing to have;
2   We must consider the afflicted person to have been unlucky; and
3   The condition can potentially be medically treated. (see Cooper, 2002, 271)

The first criterion states that in order for a patient to have a disease, the condition must be bad *for that patient* (as opposed to society at large). There are many examples of one and the same condition being good for one individual, and bad for another. Having an unusually fast metabolism, for example, is harmful to the bodybuilder (who wishes to increase muscle mass), but beneficial to the model (who can, as a result, eat a balanced diet, while maintaining a low body mass index (BMI)).[2] Cooper's claim thus implies the somewhat counterintuitive conclusion that the same condition can be a disease for one person, and not for another. That said, her example of 'sterility' goes some way to justify this – some choose to have vasectomies, yet others, who desperately want children, are sterile for reasons beyond their control. Only the latter group are diseased.

The second criterion states that an individual can be diseased only if they are 'unlucky as judged by the uninformed layman, that is, roughly, worse off than the majority of humans of the same sex and age' (2002, 276). She claims that this criterion helps us understand why one can attribute states of health to those with disabilities, and to those with genetic conditions such as Down syndrome. Down syndrome patients, she argues, are healthy 'in a certain sense,' just in case they do not have 'some other' infection or injury (276).

The third criterion, that the condition must be potentially medically treatable, separates diseases from other harmful and unfortunate states; for example, being robbed at gunpoint, losing one's job, and so on.

Objections to Cooper's model

Although the three conditions Cooper proposes do seem to apply to most diseases, the theory is subject to a variety of counterexamples. First, it is questionable whether all three necessary conditions are, in fact, necessary. I refer in particular to the second criterion, in which she states that one must be 'worse off than the majority of humans of the same sex and age' in order to be diseased. Cooper suggests that '90-year-olds who can't walk as far as when they were younger are not

DOI: 10.1057/9781137552921.0007

diseased because we expect old people to become increasingly frail' (276). This seems fair. However, there is a clear correlation between age and dementia; in an ageing population, it is worryingly possible that there will soon be an age group in which dementia is the norm. Anyone who knows somebody with dementia will confirm that the condition should always be deemed pathological, yet Cooper's conception implies that, for the very elderly, it might not be. This second criterion is at best dubious, but there are more serious problems with the thesis. Cooper presents her view as a set of necessary and sufficient conditions, yet there are numerous non-pathological states that satisfy all three.

Cooper presents and refutes a counterexample of her own – that of unwanted pregnancy. She argues that we consider this a counterexample, only because our intuitions tend to lag behind the facts. Becoming pregnant can now (assuming one is taking contraceptive measures) be 'unlucky', so unwanted pregnancy *is* a disease. The implication is that, in time, we shall appreciate this.

Whether her response is satisfactory is questionable, but I shan't dwell on this example, since one can imagine far more robust counterexamples without difficulty. I outline two possibilities below, but there are many more.

First, consider the pre-operative transsexual. A pre-operative transsexual feels he or she was either 'a man born in a woman's body', or vice versa (unlucky and bad), and one can now have a sex change (medically treatable). A pre-operative transsexual is thus, according to Cooper's model, diseased. But of course, one does not consider the transsexual to be diseased any more than the homosexual. He or she is perfectly physically able in every respect, and the transsexual is not mentally ill.[3] Perhaps Cooper might respond that our intuitions are lagging behind relatively recent developments in medicine, and that soon, now that the condition is medically treatable, we will consider the 'condition' of 'pre-operative transsexual' a disease – but I suspect not.

Second, suppose Sam decides to get a tattoo reading 'peace and love' on her arm, and, as is the fashion, decides to have it written in Chinese characters. This is explained to the tattoo artist, whom Sam knows to be very skilled, honest, and reliable. On this occasion, however, the tattoo artist gets confused, and instead of 'peace and love', Sam finds herself wandering around with 'I hate China' decorating her upper arm (albeit in aesthetically pleasing symbols). Given that Sam is about to go to

China, this is definitely bad and unlucky, but fortunately for her, this is also medically treatable. Nonetheless, it is not a disease. Cooper might respond that the tattoo *is*, in fact, pathological, but she would be clutching at straws.

Let us assume that neither Sam nor the pre-operative transsexual is diseased. Why do we take this to be the case? What strikes me is that neither Sam nor the transsexual has damaged or poorly functioning physiological traits (furthermore, neither is mentally ill in any respect). On the face of it, a plausible conception of disease in clinical medicine must accommodate this intuition.

The pure statistical conception

In this section I consider whether a value-free conception of disease will resolve the problems with Cooper's tripartite account. Although the statistical thesis does rule out cases like unfortunate tattoos and the pre-operative transsexual, the view ultimately fails (and for countless reasons). However, the basic model plays a big role in both Schwartz's account of disease (to be discussed shortly), and in Boorse's (1977) biostatistical theory (BST),[4] so a detailed exposition is warranted nonetheless.

Given the qualities expected of naturalist (value-free) views, and in particular, the thought that scientific theories often involve statistical analysis,[5] it is unsurprising that some have tried to differentiate between health and disease states using mathematical models. The pure statistical model focuses on the quantifiable properties of an organism's physiological subsystems. The precise qualities to be measured differ depending on the subsystem in question, of course, but the underlying method is (roughly) the same.

The statistical conception involves gathering data over large populations and applying formal methods to determine disease-status. Often (but not always), the measured values of a characteristic (e.g. quantities of a hormone or cell count; sizes and shapes of organs/tissues/cells; blood pressure, etc.) produce a normal distribution curve (see Figure 1.1). According to the basic statistical model, some physiological subsystem is dysfunctioning, and disease-status is met, when the measured value of the characteristic under investigation is beyond two standard deviations from the mean (Ananth 2008).

DOI: 10.1057/9781137552921.0007

One might think this gives the thesis a normative quality, since the precise value at which 'the line is drawn' (two standard deviations from the mean) seems somewhat arbitrary; that is, nothing in nature 'fixes' this value. Perhaps surprisingly, however, this is not particularly unscientific – at least insofar as it is not inconsistent with standard practice in other disciplines widely regarded as sciences. Central to testing significance in psychological and epidemiological studies, for example, is the Null Hypothesis Significance Test (NHST).[6] When one attempts to establish a non-chancy relationship between two variables, X and Y, the null hypothesis is that any correlation the data shows is accidental. The NHST states that a result is significant only if P, the probability of the null hypothesis being true (given the results of the study), is less than 0.05. This value, although not entirely arbitrary (insofar as it would not be sensible to fix the value at 0.4 or 0.8 etc.), was ultimately chosen by R.A. Fisher early in the 20th century (1925), not by nature.

The NHST is deemed a scientific method, yet the 'line-drawing' it involves is no less arbitrary than that used in the 'pure statistical method' of picking out disease states. If it is normal within the sciences to choose such arbitrary values, then there is no reason to think doing so should fall outside the remit of naturalistic conceptions of disease; that is, conceptions committed to employing only scientific methods.

Figure 1.1 shows a normal distribution curve plotting the distribution of values of a property of some physiological subsystem (e.g. the quantity of a hormone produced by an organ). The dotted lines represent standard deviations from the mean (those closest to the y-axis represent the first standard deviation from the mean, and those furthest away represent the second standard deviation from the mean). The mean, which lies on the y-axis, shows the average value of a trait across the population. The shaded sections thus represent those members of the population outside

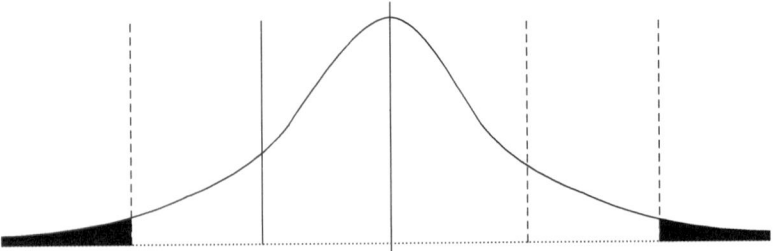

FIGURE 1.1   *The pure statistical model*

DOI: 10.1057/9781137552921.0007

of two standard deviations from the mean value. Those members of the population falling under the shaded section, then, are diseased relating to whatever trait the curve represents; that is, the trait in question is dysfunctioning.

One can see that the vast majority of the population's members fall within two standard deviations of the mean (only 4.55% are outside the two standard deviations). That extremities indicate disease is, in some cases, very intuitive – if one's BMI is far lower or far higher than the statistical norm, then one is likely to have either an eating disorder, or some other disease affecting one's BMI. Furthermore, given that the model is grounded by the quantification of physiological traits, in this way, it is clear that neither Sam with her tattoo, nor the pre-operative transsexual, is diseased according to this model. Nonetheless, the pure statistical model faces insurmountable objections.

Objections to the pure statistical model

The theory fails for a number of reasons. To mention just a few:

1 Schwartz (2007) has noted that, at least intuitively, one can identify 'healthy' and 'unhealthy' populations – we would surely want to say that, given the (on average) enormous improvements in nutrition and medical care, we (humankind) are significantly healthier in the 21st century than we were in the 16th; but of course, if one takes all and only the outliers (beyond two standard deviations) to be diseased, exactly the same proportion of the population will be healthy at any given time[7];

2 There are many unusual physiological traits that are not diseases. For example, one might have significantly more (or significantly fewer) benign moles than most people, or one may have an extra toe, or a third nipple, or unusual hair colour;

3 Some unusual physiological traits are beneficial. For example, one may have an exceptionally good immune system, or be athletically unusually able (Ananth 2008). One would not consider Mo Farah diseased because his physique is such that he can run faster and further than most people;

4 There are many common diseases – diseases that far more than 4.55% of the population suffer from (a recent study showed that 47.2% of adults in the United States have periodontal disease (Eke et al 2012));

DOI: 10.1057/9781137552921.0007

5   On the face of it, clinical judgements depend on both the
    state descriptive and normative content of patients' condition
    (Ereshefsky, 2009, 227). Consider a woman in her early-thirties
    with small, non-cancerous, endometrial polyps (non-cancerous
    growths on the inside of the uterus). One possible symptom of
    polyps is infertility. If a patient is rendered infertile by endometrial
    polyps, but is otherwise symptomless (and will most likely remain
    so), then the statistical model will deem her diseased – but
    whether she requires medical treatment will depend primarily on
    whether she (ever) wants to get pregnant. If she does then surgical
    treatment is necessary. If not, then no surgery is required. The
    clinical decision thus depends both on the state description: the
    physiological state of the patient; and a normative claim: whether
    the patient wishes to get pregnant.

In light of all these objections, one must reject the pure statistical view
as a thesis in itself. In the following section, however, I outline a more
plausible concept of disease – one that uses aspects of this statistical
approach, while avoiding its primary flaws.

The frequency and negative consequences approach, and the line-drawing problem

The problems with the pure statistical conception arise primarily from
the same underlying issue. According to the statistical model, the line
that distinguishes healthy states from diseased states is fixed at two
standard deviations from the mean. In practice, were medical records
to be analysed in this way, one would discover that the location of the
line differs from one disease to another – it depends upon the proportion of the population suffering from the negative consequences of the
condition. The bad effects of tooth decay are experienced by a large
proportion of the population, so the line is considerably further to the
right than the statistical model would dictate. However, hypertrichosis
(hair growth so excessive that it is deemed pathological – sometimes
known as werewolf syndrome) is rare. On a curve representing quantities of body hair, since fewer than 2.275% of the population suffer from
hypertrichoses, the line would be drawn further than two standard
deviations away from the mean.

DOI: 10.1057/9781137552921.0007

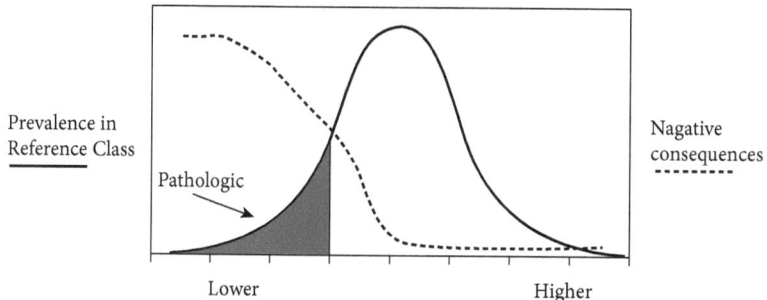

Lower   Higher

FIGURE 1.2   *The frequency and negative consequences approach*

Source: With kind permission from The University of Chicago Press. Schwartz, P.H. Defining Dysfunction: Natural Selection, Design, and Drawing a Line, *Philosophy of Science*, (July 2007): p.378, figure 2.a. ©The Philosophy of Science Association. All rights reserved.

Schwartz (2007) has proposed a solution to the problem – the frequency and negative consequences approach (FNC). According to this approach, the location of the line varies depending on how serious and how common the negative consequences are.

Schwartz introduces an additional dimension the statistical model – one representing the negative consequences of physiological conditions (Figure 1.2). The 'negative consequences' line in Schwartz's model crosses the 'prevalence in reference class' line, where the negative consequences experienced by the patient are deemed minimally 'negative' to be pathological.

Schwartz's model deals with most of the objections to the pure statistical theory: (i) the healthy/unhealthy populations objection is bypassed, since a smaller proportion of today's population will satisfy the minimally negative consequences for most diseases, than would have done in the 16th century; with respect to (ii) and (iii), none of the unusual physiological traits that are not diseases (such as red hair, or being spectacularly fit) will have the negative consequences necessary to be classified as pathological under the FNC; and the common diseases problem is avoided, since the 'negative consequences' line will cross the 'prevalence in reference class' line relatively close to the mean.

Although prima facie the addition of the 'negative consequences' dimension gives the FNC a normative dimension, this is (according to Schwartz) not the case. He takes negative consequences to be 'effects that significantly diminish the [ability to] carry out an activity that is generally standard in the species and has been for a long period of time'

DOI: 10.1057/9781137552921.0007

(Schwartz, 2007, 379), and this, he claims, allows one to differentiate between functioning and dysfunctioning 'without problematic teleological assumptions or value-judgements' (384). If this is the case, then (unlike with Cooper's heavily value-laden conception of disease) two individuals cannot be in the same physiological state, and differ with respect to their disease-status – if infertility due to polyps is a disease for one, it is a disease for all.

Objections to the frequency and negative consequences views

The FNC approach successfully dodges the line-drawing problem, but if the view is value-free, as Schwartz intends,[8] then it is unsuitable for clinical decision making. This is not a valid criticism, since Schwartz does not present the FNC approach as the clinician's concept of disease, but as an analysis of 'pathological condition'. I present the view here, rather than in Chapter 2, only because his response to the line-drawing problem (or something very much like it) is, I believe, necessary to accommodate the clinician's concept of disease.

The aim of this chapter is to identify a conception of disease suitable for clinical medicine, and one cannot make decisions about legal responsibility, eligibility for disability allowance or sick pay, and so on, without making value-judgements. Although Cooper's tripartite account of disease ultimately fails, the concept of disease she advocated was 'personalised', and this is a quality any successful conceptual analysis of disease in clinical medicine must have. In the next section I put forward an account of disease as harmful dysfunction (Wakefield 1992) – a view which, as the name suggests, includes both a naturalist (value-free) and a normativist component.

The etiological account of natural function, and disease as harmful dysfunctions

The question of what a natural function is, and indeed, whether there are such things, has been addressed extensively in the literature (Millikan 1984, 1989; Wright 1976; Ruse 1973; Neander 1991). Accounts of natural function can, rather crudely, be divided into two: the etiological and the non-etiological. Neither the statistical approach nor the FNC incorporate any kind of etiological notion into their analysis of disease. According to these conceptions, whether one is in a disease state is at least partially,

DOI: 10.1057/9781137552921.0007

if not wholly (in the case of the pure statistical account), based on the functional efficiency of one's organs, tissues, and cells, relative to the functional efficiency of those of the population. Etiological accounts of function, on the other hand, do not consider statistical notions relevant to natural function. The natural function of a trait, for the etiological theorist, depends upon its evolutionary history.

Both Wakefield's conception of disease as harmful dysfunction, and the account of disease in clinical medicine that I shall propose, require an etiological account of function.

The etiological account of function

It is generally accepted that the human heart was not *intentionally* designed to pump blood around the body (that is, by some higher power/rational being), and that, after millions of years of evolution, the best explanation for the presence of hearts is that they do, in fact, perform this function (Wright 1976; Ruse 1973; Neander 1991; Wakefield 1992). That the best explanation for the presence of hearts is the function they perform, naturally leads to an account of proper (natural) function: The function of a trait is whatever effect that trait was selected for.

To illustrate, consider Neander's example of the opposable thumb:

> it is the function of your opposable thumb to assist in grasping objects, because it is this which opposable thumbs contributed to the inclusive fitness of your ancestors, and which caused the underlying genotype, of which opposable thumbs are the phenotypic expression, to increase proportionally in the gene pool. In brief, grasping objects was what the trait was selected for, and that is why it is the function of your thumb to help you to grasp objects. (1991, 174)

In short, the etiological theory states that if a trait of a particular species has been maintained during the evolutionary process, one should take the role it plays for that species (*qua* the reason the trait has been maintained in the evolutionary process) to be its function.

Objections to the etiological account of function

Following Ananth (2008), in this section I outline and refute two objections to the type of etiological account defended by Neander (which Ananth labels 'Etiological Functional Naturalism' (2008, 176)).

DOI: 10.1057/9781137552921.0007

Objection (i)

> [I]magine a possible world in which the human lungs successfully [acclimatise] to an atmosphere that has suddenly changed from being oxygen-rich to almost entirely carbon dioxide-rich. The function of lungs before the atmospheric shift was to inhale oxygen and exhale carbon dioxide, but in the new environment the lungs function to inhale and exhale carbon dioxide. According to the etiological account, this new activity of the lungs cannot be a genuine function of the lungs because it is not the production of natural selection. The point of this counterfactual is that the etiological account is forced to accept the counterintuitive view that many beneficial acclimations that help keep an organism alive are not functions. (2008, 180)

Response to objection (i)

It may well be the case that the etiological account of function implies that some beneficial acclimatisations are not functions, but whether this is overly problematic is not so obvious. Arguments that appeal to intuition are questionable – if nothing else, because intuitions often differ between parties. It is *my* intuition that in the example above, if the inhaling and exhaling carbon dioxide-mechanism is not yet a part of a 'natural selection history', then it is not (yet) a natural function of the lungs. That it is beneficial to survival and reproduction (in Boorse's sense) in the present circumstances is not sufficient for 'natural function' status.[9]

Objection (ii)

Human beings have a number of vestigial physiological subsystems and stimulus-response mechanisms, the traditional example being the appendix,[10] but goose bumps, the vomeronasal organ, and several others fall into this category, too. The etiological account of function, however, seems to imply that these organs must still have a natural function[11]: that which explains their existence.

Response to objection (ii)

One *might* think that Neander's approach entails that the appendix has a function: to digest plant-based foods with very little nutritional value. The appendix ceased to perform this function some time ago (since we no longer eat these low-nutrition foods), so it no longer contributes towards our natural goals, but nonetheless, helping with digestion might still be deemed the natural function of the appendix.

DOI: 10.1057/9781137552921.0007

Unfortunately, this is of little help to those wishing to incorporate an etiological conception of natural function into an analysis of disease. If one grants that all so called vestigial organs have a function, all vestigial organs are always diseased, since none perform whatever function is assigned. The only option is to deny that vestigial organs have a function, despite there being an evolutionary explanation for their existence. This is not, however, as problematic as it first seems – one can add a further condition that resolves the issue. I suggest one adapts the etiological account as follows:

1   A physiological subsystem has a natural function only if it contributes to the organism's natural goals.
2   If a physiological subsystem has a natural function, then it is that which explains its existence in the organism.

One possible worry here is that any view that appeals to dysfunction in its conceptual analysis of disease will not be able to accommodate diseases of vestigial organs (such as appendicitis). On closer inspection, however, this is not so problematic. Vestigial organs cannot dysfunction, but they can cause other traits to dysfunction; in other words, although a vestigial organ like the appendix can cause a disease, the organism is diseased in virtue of the non-vestigial organ dysfunctioning.

Wakefield and diseases as harmful dysfunctions

Wakefield outlines his etiological account of disease as follows:

> A condition is a disorder if and only if (a) the condition causes some harm or deprivation of benefit to the person as judged by the standards of the person's culture (the value criterion), and (b) the condition results from the inability of some internal mechanism to perform its natural function, wherein a natural function is an effect that is part of the evolutionary explanation of the existence and structure of the mechanism (the explanatory criterion). (1992, 384)

The value criterion permits this concept of disease to be used by the clinician, since it accommodates the fact that two people with the same physiological condition can differ with respect to disease status. The explanatory criterion ensures that the counterexamples to Cooper's conception are not applicable – Sam's tattoo, although medically treatable, is not the consequence of some internal mechanism failing to perform its natural function (and so does not qualify as a disease), nor is the pre-operative transsexual's – so far so good.

DOI: 10.1057/9781137552921.0007

Objections to Wakefield's disease as harmful dysfunction

Wakefield's approach, as it stands, fails to deal with Schwartz's line-drawing problem. There is a range of healthy levels of functional efficiency. One's blood pressure is healthy just in case one's systolic blood pressure is below 140 and above 90, and one's diastolic blood pressure is below 90 and above 60. Wakefield claims that his etiological approach can accommodate this – the healthy levels of functional efficiency are distinguishable from the pathological, since those in the healthy range were favoured by natural selection. However, Schwartz writes:

> [Wakefield's] approach does not solve the line-drawing problem. The first difficulty is that for acquired disorders, ones not caused genetically, natural selection does not apply. For example, it is not the case that Mr. Smith's [ejection fraction] of 20% is favoured by natural selection or not favoured, since his condition is not hereditary. (2007, 370)

On the face of it, Wakefield's account of disease as harmful dysfunction is well-suited to the clinician, since it permits medicine to be personalised. However, doctors also need to distinguish the healthy from the pathological levels of functional efficiency, and as it stands, Wakefield's approach fails in this respect.

Disease as harmful function – 'drawing the line' on the etiological account of disease

Drawing on both Schwartz and Wakefield, in this section I propose new and feasible concept of disease in clinical medicine – disease as harmful function.

Although Schwartz resolves the line-drawing problem, his view is not suitable for clinical medicine. Wakefield's approach, on the other hand, is *prima facie* a good candidate. However, Schwartz's response to the line-drawing problem is not applicable to Wakefield's conception, and no alternative response is provided. Wakefield's harmful dysfunction account thus includes the evaluative component necessary for the clinician, but it cannot provide a naturalistic account of *healthy* functioning; that is to say, although an etiological account of function can identify the natural function of a trait, it cannot identify the associated range of 'healthy' functional efficiencies.

DOI: 10.1057/9781137552921.0007

Schwartz's argument is compelling, since there are many disorders for which Wakefield's proposed response fails. However, Schwartz's response to Wakefield does not rule out an etiological approach *tout court*.

As we have seen, the etiological account of natural function identifies the natural function of a trait as the evolutionary explanation for that trait's existence. Identifying the natural function, then, is not problematic for the etiological theorist. Providing a value-free way of identifying the healthy range of efficiencies would be, but fortunately, she doesn't have to.

Any plausible account of disease in clinical medicine is value laden, and indeed, the first clause of Wakefield's conception: '[a disease] causes some harm or deprivation of benefit to the person as judged by the standards of the person's culture (the value criterion)' (1992, 384) is inherently so. The second clause, that '[the harm] results from the inability of some internal mechanism to perform its natural function' (384) is problematic, for it forces Wakefield to provide an account of dysfunction. An account of dysfunction, however, can be replaced with a list of trait-values (e.g. BMI values, blood pressure values, quantities of insulin produced, etc.), plus the degree of harm/benefit associated with those values (Figure 1.3).

The etiological theorist can now offer an account of disease as harmful *function*. A condition is a disorder if and only if '(a) the condition causes some harm or deprivation of benefit to the person as judged by the standards of the person's culture (the value criterion), and' (b) the condition results from an internal mechanism performing its natural function, at a harm-causing level of efficiency (that is, a level of efficiency that is *bad for the individual*), 'wherein a natural function is an effect that is part of the evolutionary explanation of the existence and structure of the mechanism (the explanatory criterion)' (384).

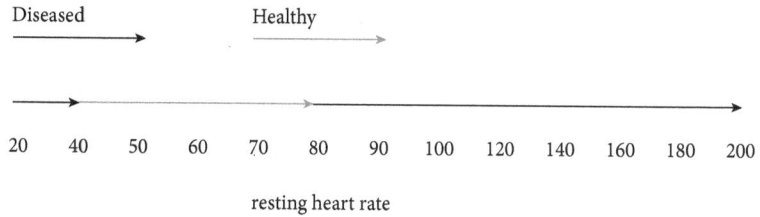

FIGURE 1.3   *Disease as harmful function*

DOI: 10.1057/9781137552921.0007

In Figure 1.3 the range of healthy resting heart rates are represented by the grey arrow, and all other resting heart rates are represented by the two black arrows. The function of the heart is to pump blood around the body. Most of our hearts do pump blood around the body within that healthy range, but according to the harmful functioning conception, that is not what determines the healthy range. The range of healthy resting heart rates (where the lines are drawn) is chosen by medicine. If a patient's resting heart rate does not fall within the healthy range; that is, if it is deemed to be at level of efficiency bad for the patient, then according to the clinician, it is diseased.

Conclusions

Cooper's account of disease is heavily value-laden, and therefore equipped to deal with the evaluative aspects of clinical medicine. However, this tripartite account is subject to many counterexamples. One of the necessary conditions implies that dementia may not be a disease for some, and there are numerous 'conditions' that satisfy all of her criteria, but are clearly not diseases (e.g. having unfortunate tattoos).

The pure statistical view is value-free, and hence unfit for the clinician's conception of disease. Furthermore it is subject to counterexamples that render the view entirely implausible as a stand-alone conception of disease. However, Schwartz incorporates the statistical approach into his FNC approach, and although (as a value-free account) it is not suitable for the clinician's concept of disease, we see in Chapter 2 that it is a plausible candidate for the pathologist's.

Wakefield's disease as harmful dysfunction account fails because it cannot provide a satisfactory response to the line-drawing problem; that is, it cannot draw the line between where health ends, and dysfunction begins. 'Disease as harmful function', however, is value-laden to the extent required by the clinician; it ensures that diseases involve harm caused by 'internal mechanisms' functioning in an undesirable way; and this is not affected by its inability to provide a naturalistic account of dysfunction, since dysfunction does not feature in the view. Disease as harmful function, then, is a good fit for the clinician's concept of disease.

DOI: 10.1057/9781137552921.0007

Notes

1   Consider the emotion 'despondency' – should a despondent patient automatically qualify for treatment with antidepressants? The decision medical practitioners make concerning treatment, may hinge on whether 'despondency' (*simpliciter*) is classified as a disease.

2   Here I assume that the model's weight is not harmful to her in other respects, e.g. underweight to the extent that her organs function poorly.

3   That is assuming, of course, that he or she is neither mentally nor physically ill in some other respect (e.g. chest infection, schizophrenia, etc.).

4   See Chapter 2 for a detailed discussion of the BST.

5   This is demonstrably true in the medical sciences. The results of epidemiological studies are almost always grounded by statistical methods.

6   It is worth noting that this method has not gone unchallenged (Kirk 1996), but nonetheless it is still frequently used.

7   Kingma (2010) presents a similar argument, noting that in 'harmful situations', the vast majority of the population can become diseased (e.g. universal exposure to a dangerous pathogen) – but this inference is incompatible with the statistical theory, since the level of functional efficiency required to be healthy would drop such that precisely the same proportion of the population were diseased.

8   Kingma has argued that Schwartz may have failed in this respect (2014, 596).

9   See Ananth (2008, 181) for further discussion.

10   Whether or not the appendix is vestigial is, in fact, questionable. Matsushita, Uchida, and Okazaki (2007), for example, conducted a study suggesting that it plays a significant role in the pathogenesis of ulcerative colitis.

11   See Ananth (2008, 181–182).

DOI: 10.1057/9781137552921.0007

2
What is a Pathological Condition?

Abstract: *This chapter is a conceptual analysis of the theoretical notion of disease, in which I ask: 'what is a pathological condition?' The focus of this chapter is by far the most frequently cited 'naturalist' conception of disease/ pathological condition: that proposed and developed by Boorse. I go on to show that by replacing the biostatistical conception of natural function, with Karen Neander's etiological account, one can espouse a conception of a pathological condition immune to all the BST's problems.*

Keywords: Christopher Boorse; etiological theory of pathological condition; naturalism; the biostatistical theory; the concept of disease

Smart, Benjamin. *Concepts and Causes in the Philosophy of Disease.* Basingstoke: Palgrave Macmillan, 2016. DOI: 10.1057/9781137552921.0008.

DOI: 10.1057/9781137552921.0008

Introduction

The target of Chapter 1 was a conceptual analysis of disease for the clinician. Here I analyse the concept of disease from a slightly different perspective. This chapter is a conceptual analysis of the *theoretical* notion of disease, in which I ask: 'what is a pathological condition?'[1]

Although the clinician's concept of disease is clearly inherently value-laden, this is not true of pathological conditions. The focus of this chapter is by far the most frequently cited 'naturalist' conception of disease/pathological condition: that proposed and developed by Boorse (1975; 1976; 1977; 1997; 2014). I ultimately argue that the biostatistical theory (BST), as it stands, has not been rescued from Guerrero's (2010) Cambridge change objection. Furthermore, I show that by replacing the biostatistical conception of natural function, with the etiological account of function outlined in Chapter 1, one can espouse a conception of a pathological condition immune to all objections raised against the BST.

What is naturalism about disease?

Before outlining the BST, it is worth emphasising Boorse's naturalist target. This is especially important, since many of the objections to his thesis arise from confusion in this regard. For example, Murphy writes:

> The naturalist conception of disease (perhaps most clearly stated in Boorse 1975, 1997) is that the human body comprises organ systems that have natural functions from which they can depart in many ways. Some of these departures from normal functioning are harmless or beneficial, but others are not. The latter are 'diseases'. So to call something a disease involves both a claim about the abnormal functioning of some bodily system and a judgement that the resulting abnormality is a bad one. (2015, 2)

Although Murphy is right to state that the naturalist conception (or at least Boorse's naturalism) concerns departures from the normal functioning of physiological subsystems, the claim that the naturalist concept of disease involves a normative judgement is somewhat misguided. Indeed, this is precisely what Boorse denies. He takes disease to be 'a type of internal state which impairs health' (Boorse, 2014, 684) – but it is not value-laden, since none of the ideas comprising the concept of

DOI: 10.1057/9781137552921.0008

pathological condition are value-laden.[2] Boorse's target is to 'explain key concepts of biology and medicine' (695) – no more, no less.

Boorse's naturalism

According to Boorse, physiological subsystems (organs, tissues, cells) have a natural goal: to contribute to survival and reproduction. A subsystem is diseased, he says, when its contribution to survival and reproduction is subnormal for the organism's reference class; that is that relative to the biostatistical norm for the organism's age, sex, and species, the subsystem makes a poor contribution to its natural goals.

Boorse outlines the BST as follows:

1 The *reference class* is a natural class of organisms of uniform functional design, specifically, an age group or sex of a species.
2 A *normal function* of a part or process within members of the reference class is a statistically typical contribution by it to their *individual* survival and reproduction.[3]
3 A *disease* is a type of internal state which is either an impairment of normal functional ability, i.e. a reduction of one or more functional abilities below typical efficiency, or a limitation on functional ability caused by environmental agents.
4 *Health* is the absence of disease. (Boorse, 1997, 7–8)

The resultant picture resembles that of the pure statistical conception discussed in Chapter 1, with the important exception that only subnormally functioning traits can be diseased.

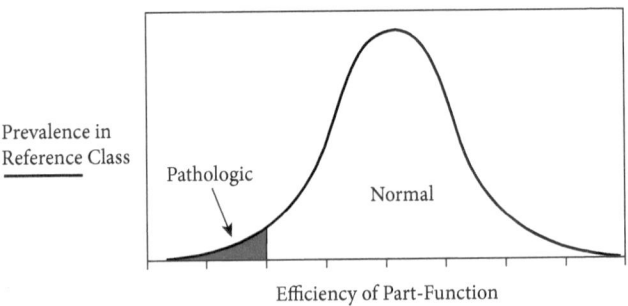

FIGURE 2.1   *The biostatistical theory*

Source: With kind permission from The University of Chicago Press. Schwartz, P. H. Defining Dysfunction: Natural Selection, Design, and Drawing a Line. *Philosophy of Science* (July 2007): 272. figure 1 © The Philosophy of Science Association. All rights reserved.

DOI: 10.1057/9781137552921.0008

At this point, it is worth highlighting a few subtleties. First, Boorse's reference classes are not species *simpliciter*, but subclasses of species (determined by age groups and sex). This prevents prepubescent children and postmenopausal women from being diseased due to infertility; men's testes, liver, and adrenal glands being diseased for creating subnormal (for the species) quantities of oestrogen (despite creating the mean level of oestrogen for their sex), and many hundreds of similar examples.

Second, one possible interpretation of the second clause is: 'A *normal function* of a part or process within members of the reference class is a statistically typical contribution by it to the survival of the individual organism, *and to the ability of the organism to have offspring*,' since with respect to the former Boorse refers specifically to the organs and processes of the individual organism, and there is little indication that he moves from speaking of individual organisms to, say, their genes (or indeed to anything else). Boorse later indicates, however, that 'homosexuality' may or may not be a disease – its status depending on whether or not one of the kin-selection hypotheses is true (2014, 691). Homosexuality is undoubtedly detrimental to an organism's propensity to produce their own offspring, so one would think that the BST points towards it being classified as a disease[4]; but if a kin-selection hypothesis is true, homosexuality is not detrimental to the reproduction of the homosexual's *genes*, since the individual's genes are shared by his sibling's children (children whose survival is *ex hypothesi* aided by the homosexual). If the disease status of homosexuality rests on the truth of one of the kin-selection hypotheses being true, then Boorse's natural goal of reproduction refers to the subsystems' contribution to the reproduction of genes, and not to the individual organism's ability to have offspring. This is just as well, because Boorse intends the BST to be applicable across the natural world (that is, not only for humans, but also for non-human animals and plant life), and there are many cases in which individual organisms cannot, by natural selection, themselves have offspring. Worker bees and worker ants, for example, labour to ensure the continuation and growth of their colony, but even when perfectly healthy, they cannot reproduce – at least not in the way the 'naïve' reading of clause two suggests.[5]

This schema correctly identifies many diseases. One species-typical function of the pancreas (more specifically, the beta cells within the pancreas), for example, is to produce appropriate quantities of insulin to regulate blood-sugar levels, preventing hyperglycaemia. This contributes to the organism's natural goals, since it helps prevent myocardial infarction, kidney failure, and stroke (all of which are detrimental to individual survival and reproduction!). According to Boorse, if the pancreas is producing subnormal quantities of insulin, then the organ is diseased, and of course, medicine agrees.[6]

One can also see how this analysis does not fall foul of at least some of the objections to the 'pure' statistical conception. The abnormal qualities of Farah (or at least those contributing to his athleticism) do not impair the functional efficiency of his biological subsystems. Indeed, the functional efficiency of his muscles and lungs is clearly significantly higher than most, since in a situation where running is necessary (say, from prey), their contribution towards his natural goals is unusually good. Given that Farah's lungs are not functioning subnormally, they are not diseased. A similar response applies to the 'third nipple', and 'unusually few benign moles' examples. They do not reduce functionality relating to the organism's natural goals, and thus do not count as diseases, either. Furthermore, the second disjunct of the third clause accommodates (at least some) cases where disease is the statistical norm, avoiding the usual counterexamples such as gum disease (Boorse 1997).

Objections to Boorse's naturalism

In the following sections I present a number of existing objections to Boorse's naturalism, some more compelling than others. Most, however, can be responded to without too many concessions.

Objection (i) *The thesis is both too strong, and too weak*

Boorse's conception *prima facie* looks both too strong, and too weak. Too strong, since as Joseph Margolis has pointed out:

> [It] is quite possible to imagine a set of circumstances in which eyes would lose their function and yet not be diseased. For example, imagine that, because of terrestrial pollution, the human race adopts, and adapts to, a life maintained at a submarine level unpenetrated by water. (1976, 247)

Similarly, one can choose to make one's biological subsystems perform subnormally/malfunction/dysfunction, and subsequently choose to restore their functional efficiency, for example, when one uses earplugs, wears a blindfold, or dives underwater (Flew 1973). Yet clearly one is not diseased when the visual impairment is intentionally caused by an artificial, external item, which can be removed at will.

DOI: 10.1057/9781137552921.0008

The thesis is too weak, since one can be diseased without subsystems malfunctioning. For example, a patient may have had multiple epileptic seizures, but thanks to effective medication be seizure-free for several years. It would be odd to claim this patient does not have a disease (and one requiring treatment, at that), when, if the patient stopped taking the medication, the seizures would return.

Response (i)

These objections can be dealt with fairly easily. First, the strange scenario Margolis outlines is unproblematic for Boorse, since the species-typical contribution of the eyes to individual humans' reproduction and survival would be negligible in those circumstances; second, Boorse can accommodate blindfolds and earplugs, since the focus of the BST is not actual subnormal functioning, but the disposition to dysfunction. . He quite clearly states that:

> [In my 1977 I required] 'functional readiness', not just current function, and noted that 'biological functions are usually performed on appropriate occasions, not continuously'. (2014, 685)

If one has healthy eyes but is blindfolded, one retains the disposition to see when the blindfold is removed; and even while taking the epilepsy medication, an epileptic is disposed to have seizures under normal conditions (that is, when no medication is being taken).

Objection (ii) There are many possible reference classes

Boorse's reference class criterion is vital to the success of the biostatistical theory, but Cooper has argued that this clause is problematic, not because separating into reference classes is unnecessary, but because separating into sex, age, and species-type is insufficient. Empirical evidence, she argues, suggests that if one were to adopt an account of disease of this nature, considerably more reference classes than Boorse proposes are necessary.

> What's normal for an organism depends not only on species, sex and age, but also on a host of other factors. Masai are naturally sensitive to growth hormone, pygmies are not [, and athletes] normally have a lower heart rate than other people. (Cooper, 2002, 266).

To accommodate the lower levels of oxygen in the air, those who live at high altitudes (on average) have a higher red blood cell count than those who live at lower altitudes; so what counts as a subnormal red blood cell

DOI: 10.1057/9781137552921.0008

count should, it seems, be lower for those living at lower altitudes. If one grants this *prima facie* intuitive conclusion, then the reference classes must include more categories than just those of age, sex, and species. Cooper argues one must include (at least) race, environment, and training, as well.[7]

Of course, introducing additional reference classes reduces the average population-size. If the classes become too small, then the thesis becomes implausible, since the smaller the population-size, the less likely it is that the distribution curve will cohere with the medical consensus on which physiological states are pathological. Cooper even suggests that the population of some reference classes could be just one, in which case, every member of that population (i.e. the only member) *must* be healthy in every respect, for she, alone, determines the mean levels of efficiency.

Response (ii)

Rather than attacking the 'many references classes' objection from an 'in principle' perspective, Boorse (2014, 702–705) takes each of the classes proposed by Cooper and disputes them individually.

Cooper suggests that race must be a reference class – but Boorse has already argued in his 1997 that it *may* be necessary to add race to his list. Adding race would not be overly problematic, since race (alone) would not reduce the size of the reference classes significantly enough for Cooper's objection to hold any weight.

Regarding environmental factors, the response to objection (i) is equally applicable; if one is healthy, one's physiological subsystems are disposed to perform their functions with (at least) species-typical efficiency in typical circumstances. Those who have adapted to live at high altitudes have a higher red blood cell count than those living at low altitudes, but the circulatory systems of healthy individuals at both high and low altitudes, are *disposed* to perform their function with at least normal efficiency in typical circumstances; that is, in typical circumstances, those individuals would have at least species-typical red blood cell counts.

Finally, Boorse dismisses the physical training examples as 'a joke'. It is of course true that the typical heart rate of a professional cyclist is lower than that of a non-cyclist; but a professional cyclist with a heart rate typical for a non-cyclist does not (automatically) have a pathological condition. The species-typical heart rate is *medically normal*, so assuming the cyclist's heart is not disposed to function subnormally in typical circumstances, the cyclist is not diseased (2014, 703).

DOI: 10.1057/9781137552921.0008

By addressing each of Cooper's proposed references classes in turn, and showing why none of them (with the possible but not unproblematic exception of 'race'[8]) should be included in his schema, Boorse has undoubtedly diffused Cooper's argument. It has not been categorically proven that further references classes are unnecessary, of course. Perhaps classes other than those proposed by Cooper are required, and perhaps better counterexamples to the existing thesis exist – but this looks unlikely, and until they are provided, the BST cannot be refuted on these grounds.

Objection (iii) The biostatistical theory is value-laden

As we have seen, to avoid counterexamples such as 'children must be fertile to be healthy', Boorse introduces the reference classes of age, sex, and species-type. But what if the references classes had been chosen differently? By choosing age, sex, and species-type, rather than 'Down syndrome' and 'long-term alcohol abuse', Boorse is able to claim that Down syndrome and liver cirrhosis are diseases (Kingma 2007); however, the BST strategy would be forced to conclude otherwise were the references classes different (DeVito 2000). So what justifies Boorse's particular selection of reference classes? Although it clearly makes more sense to choose the classes of age, sex, and species-type than those of 'Down syndrome', and 'long-term alcohol abuse' (partly, perhaps, precisely because the results cohere better with those states we deem pathological), the mere act of choosing reference classes (it is argued) involves making an evaluative judgement.

The biostatistical theory also implies that if the function of some subsystem pertains to anything other than the goals of survival and reproduction, then its efficiency is irrelevant when assessing disease states. This seems wrong. Human beings have goals that reach far beyond those of survival and reproduction. One would imagine, for example, that successfully fulfilling one's goal of writing good philosophy is far less conducive to reproduction than successfully fulfilling one's goal of being a rock star, but writing good philosophy and becoming a rock star are equally legitimate goals, regardless of the respective conjugal implications. Indeed, one may actively desire *not* to reproduce. Should one deem all those who have had vasectomies to be diseased? The 'common sense' conclusion is surely 'no' – but those who have chosen to have vasectomies unquestionably have a subnormal ability to reproduce.

DOI: 10.1057/9781137552921.0008

If one does not think all those with vasectomies are diseased, then either the list of/criteria for 'natural' goals requires revision, or one needs to abandon the notion that natural goals feature in the concept of disease altogether, perhaps in favour of more general value judgements. The latter, of course, is a normativist strategy through and through. Either way, whether one chooses 'survival and reproduction', or alternative goals pertaining to more general wellbeing, purely in choosing the goals it seems one is making a value judgement. In short: one makes judgements both in choosing the 'natural' goals, and in choosing the reference classes, and a naturalist account involving evaluative judgements is no naturalist account at all.

Response (iii)

Objection (iii) is a popular one, but it is entirely misguided. Boorse sets out to provide a conceptual analysis of disease consistent with scientific method, using only the language of science – his aim being 'in some sense an attempt at a "lexical definition" (1977, 551) of scientific terms' (2014, 713).

The key point is this: for a concept to be value-laden, it must include value-laden terms or ideas. The biostatistical theory does not. Neither survival nor reproduction nor age nor sex nor species are value-laden terms. One might push that survival and reproduction form part of the concept only because they are the natural goals, and 'goal' is a value-laden concept – but that won't help. Boorse's four-part schema describes functions only in terms of the contribution of subsystems to survival and reproduction (there is no mention of goals, here) – neither the idea of survival, nor the idea of reproduction, is value-laden.

That the component parts of the disease-concept are what they are, may involve *some* kind of choice, but the concept of disease 'already exists as a target [in pathology] ... So the only way to run an argument of this type is to claim that [pathology] – not the [BST] – has chosen one of many possible health concepts' (2014, 693).[9] Medicine chose the content of the concept of disease, but only in the same sense that biology chose the criteria for being a mammal. That does not make 'mammal' a value-laden concept!

Boorse goes on to write:

> None [of my critics] gives sense to the idea of a value-based 'choice of the health concept' by [pathology]. The obvious way to do so fails. That is to assume from the outset that 'disease' is an evaluative concept, meaning

something like 'undesirable condition' or 'condition deserving medical treatment.' Then [pathology] will choose what falls under this description – *i.e.* make value judgements about what conditions are undesirable or need treatment. But, so clarified, the argument has two fatal defects. First, it is circular, since it assumes its conclusion, that 'health' is value-laden. Second, it ignores one of the most basic features of medical usage of 'disease': that *disease* and *medically treatable condition* do not coincide. (2014, 693)

Boorse must surely be right. Consider the case of the vasectomy. One may argue over whether being infertile is *always* a state of disease (whether naturally occurring or self-inflicted), but surely one would never consider being fertile a disease! Nevertheless, we can, and do, medically 'treat' fertility with vasectomies and other contraceptive methods when high fertility is an undesirable physiological state. 'Disease' and 'medically treatable condition' are altogether different concepts.

For Boorse, it is determined empirically that the goals and reference classes specified are the *right* goals and reference classes. We have already seen that there are several candidate reference classes. I have spoken of age, sex, species-type, race, training, environment, long-term alcoholics, and people with Down syndrome, but of course there are indefinitely more possibilities. For the naturalist, though, some of these are the right references classes, and the rest are not. In medicine, whether or not D is a disease is an empirical matter of fact. For mental disorders, for example, one need only look at The Diagnostic and Statistical Manual of Mental Disorders (DSM-5). If the right goals and reference classes are those that correctly identify diseases (which seems to be the case, if one is searching for a lexical definition), then discovering the right goals and reference classes is purely a matter of empirical enquiry, with no normative component.[10]

So why does the normativist think the naturalist has a problem, here? Here is one possibility: the theory 'Constructivism' in the philosophy of medicine (as presented by Murphy 2015), and the fact that the concept of disease has been, in some sense, 'constructed' (by medicine), have been conflated. Boorse's naturalist view is constructed, but only in the sense that the concept of fish has been constructed (after all, biology could have chosen 'having fins' to be sufficient for being a fish). But that there is a choice component to the concept of disease does not imply that the concept is value-laden. It does not imply that it is 'Constructivist' in Murphy's sense:

> The key [C]onstructivist contention is that there is no natural, objectively definable set of human malfunctions that cause disease. Rather, constructivists assert that to call a condition a disease is to make a judgment that someone in that condition is undergoing a specific kind of harm that we explain in terms of bodily processes. (2015, 3)

Murphy's Constructivism clearly involves a normative component, but the BST does not resemble Constructivism at all – the BST makes no reference whatsoever to patients undergoing harm. Although there is of course a 'choice' component to Boorse's concept, as there is with most concepts in the sciences (and indeed more generally), that concept involves no value-laden ideas, and it is hence not value-laden.[11]

Objection (iv) The Cambridge change objection

It is not uncommon in the medical sciences for conclusions to be reached by comparing the qualities of one reference class to another, or of one individual to another; the results of these contrastive studies depend both on the state of an individual (or of a cohort), and on the state of the reference class. Since populations are subject to change, both with respect to the members of the population, and with respect to the properties of the members, what is statistically typical is also subject to change; according to BST, then, health status is subject to change *without any physiological change in the individual.*

Suppose the body mass index (BMI) of >30 indicates disease. According to the BST, this fact is grounded by what is species-typical. The BMI level indicating disease will thus increase, if the average BMI of the population increases. Now suppose *you* have a BMI of 30, and currently qualify as diseased under the BST. Should you believe that your health improves as the species-typical BMI increases to 30, even though your own weight remains constant? Presumably not, since you are equally susceptible to myocardial infarction, diabetes, and so on. A change in disease-status without a change in intrinsic physiological qualities is known as a 'Cambridge change' (Geach 1972), and the thought that health status changes can be Cambridge changes is surely absurd (Guerrero 2010).

Response (iv)

Boorse provides a three part response:

    (1) the theoretical possibility of a Cambridge change in someone's health status seems to be entailed by any view of health as normality, an idea basic

DOI: 10.1057/9781137552921.0008

to scientific medicine. But (2) it is far more difficult than Guerrero thinks for such a change to occur, and (3) the only realistic ways in which one could occur are of no importance to medical practice. (2014, 715)

These responses are convincing if one accepts that health as normality is basic to medicine. Given that Boorse's target is a lexical definition of scientific terms, this, I suppose, is an empirical matter. However, even if it is true, one must question whether the concept of health might lose its connection to normality if any Cambridge change situations actually occurred. If one is using the Cambridge change objection to *question* conceptions of health as normality, however, response (1) is question-begging.

Regarding (2), *prima facie* Cambridge change situations are not as difficult to imagine as it might seem – the sudden outbreak of bubonic plague in the middle ages, for example, would have a significant effect on the average functional efficiency of many traits, and of course, that event was of significant importance to medical practice. Boorse has a response to this, however. The BST does not, he says, base normal health on the species-typical levels of functional efficiency at particular 'moments', but over temporally extended 'time-slices' (Boorse, 2002, 99). This means that a quick change in the mean efficiency of some subsystem will not significantly affect the health status of individuals, since those individuals at the present moment form only a small part of the population over the time-slice (in its entirety).

Suppose a meteor hits the Earth and kills all human life with the exception of one individual, who merely loses an eye. Were one to ask: 'How many functioning eyes does a healthy human have?', given Boorse's additional criterion, the answer would be two, since for a long period before this unfortunate celestial encounter (that is, during the rest of the time slice), the mean number of functioning eyes for a human being was two. The fact that, at this precise moment, the mean number of functioning eyes is one, does not affect this in any significant way. The same principle applies to rapid spreads of infection, and so on.

This response, although initially appealing, is at best incomplete. The 'time-slice' Boorse appeals to must be specified (or at the very least, more needs to be said about it), for very lengthy time-slices often produce as many counterintuitive consequences as the 'single-moment' time slices assumed by Guerrero. Most of us are happy to say that the average health of human beings, at least in western countries, has improved over the last

500 years or so; not least because improvements in nutrition, hygiene, and developments in medical care have significantly increased both life expectancy, and quality of life. On the face of it this is supported by Boorse's hypothesis, since the average species-typical contribution of our physiological subsystems to our natural goals is higher than it was in 1515 CE. However, if one takes the time-slice factor into consideration, and suppose Boorse's time-slice covers 500 years, many people now considered to be malnourished and diseased would be classified as healthy, since what is species-typical over the time-slice partly depends on the (relatively) unhealthy physiological states of those living in the 16th century. Boorse does suggest that any period shorter than 'lifetime or two' is too short for the necessary time-slice,[12] so 'one can only conclude that [Boorse thinks] the time slice is between two lifetimes and a millennium' (Giroux, 2015, 185). In light of the above, it should probably be closer to two lifetimes.

Élodie Giroux (2015), however, has recently (and reasonably) claimed that the concept of health should cohere with developments in epidemiology. After all, epidemiological research is of central importance to the prevention, diagnosis, and treatment of disease. She argues that Boorse's suggestion of between two lifetimes and a millennium is far too long for the time-slice, since it could not accommodate epidemiological studies:

> theoretical medicine and physiology cannot ignore epidemiological changes in human morbidity and in human longevity...Thus, far from supporting the idea of a stable and natural human functional design, descriptive epidemiology reveals the importance of historical evolution of health phenomena at the population level and brings to light a crucial issue: the relevant time slice extension of a reference class...it seems to me that a shorter population time-slice to which our theoretical health judgments are relative would better fit contemporary medicine. (Giroux, 2015, 186)

If the time slice is short, then Guerrero's Cambridge changes will be rife; if it is centuries, then the average functionality will be so low, that far too many present people will be deemed healthy; and if it lies somewhere in the middle, say, two lifetimes, then it is incompatible with current assumptions and progress in epidemiology and clinical medicine. A convincing response to the Cambridge change objection may appear at some point, but this objection is certainly a threat to the BST.

DOI: 10.1057/9781137552921.0008

The frequency and negative consequences approach revisited

We saw that in order to resolve the line-drawing problem, Schwartz suggests that an extra dimension to the BST should be added – that of 'negative consequences'; wherein negative consequences are 'effects that significantly diminish the ability of a part or process in the organism ... to carry out an activity that is generally standard in the species' (2007, 379).[13] This additional dimension allows the FNC approach to combat the common diseases problem, but it also resolves the Cambridge change objection. The Cambridge change objection states that the levels of functional efficiency necessary to be healthy, rises and falls with the species-typical functional efficiencies – thus one can be healthy one day, and diseased the next, despite no changes in the functional efficiency of one's own traits. This does not apply to the FNC approach, however, since the negative consequences condition moves 'the line' as the average levels of functional efficiency rise and fall (Chapter 1 of this book, pages 12–14). As the average level drops, the line moves to the right such that a higher proportion of the population are diseased; and as the average functional efficiency rises, the line moves to the left, such that a lower proportion of the population are diseased.

The FNC is promising, Kingma, however, questions the naturalism of the negative consequences factor, pointing out that 'sometimes, where the situation demands it, "the ability to carry out a standard activity" needs to be suppressed or impaired' (2014, 596)[14] – one cannot breathe and swallow at the same time, but this is not pathological. A value judgement needs to be made, then, to determine which abilities can be suppressed without affecting health status. The FNC may or may not be a purely naturalist view, but either way, it does seem to deal with the Cambridge change objection that the BST looks to struggle with.

The target of this chapter is a plausible conceptual analysis of pathological condition. I do not assume this *must* be entirely value free, however. If Schwartz's negative consequences condition is value laden, this only shows Schwartz to be mistaken about *that*. It does not imply that the FNC is implausible as a theoretical conception of disease – the FNC would not permit, for example, two individuals with the same set of trait functional efficiencies, to differ with respect to their health status.

DOI: 10.1057/9781137552921.0008

The etiological theory of pathological condition

The FNC approach is plausible if one prefers non-etiological conceptions of natural function, but otherwise not. In this final section I propose a further naturalist account of pathological condition; this view, like the BST, states that diseases are impairments to the biological functions that contribute to survival and reproduction. The view I propose – the etiological theory of pathological condition (ETPC) – draws heavily on Boorse and Schwartz, but rejects the strong link between theoretical health and biostatistical normality. Instead, the crucial second clause is replaced with an etiological account of function.

Below is the 'etiological theory of pathological condition' (. Note first that the effect for which a physiological subsystem was chosen differs between reference classes – natural selection chose that healthy prepubescent boys cannot reproduce, and natural selection chose that adult men can. Clause 2 refers to the 'reference class relative effect for which that trait was selected'. This covers the need for the natural function(s) of some traits to differ between reference classes.

1   The *reference class* is a natural class of organisms of uniform functional design, specifically, an age group or sex of a species.

2   The natural function of a trait within members of the reference class is the reference class relative effect for which that trait was selected, *unless* it fails to contribute to individual survival, or reproduction of the organism's genes. In which case that trait has no function.

3   A *disease* is a type of internal state which is (a) either an impairment of a natural function; *i.e.*, a reduction of functional ability, or a limitation on functional ability caused by environmental agents; and (b) is deemed by medicine to have sufficient negative consequences (to a level chosen by the *pathologist*) to be pathological, where negative consequences are 'effects that significantly diminish the ability of a part or process in the organism ... to carry out an activity that is generally standard in the species'.

Now let us return to the Cambridge change objection. ETPC is not threatened by Cambridge changes. It draws on our best biological theories to produce an account of function for biological-types, where the natural function of a trait is the reference class relative effect for which that trait was selected. ETPC identifies disease wherever a physiological state impairs a natural function,[15] such that it has at least the minimally negative consequences for the organism, as chosen by the pathologist. This makes no reference to biostatistical normality – functions are defined etiologically, and health status is determined only by internal physiological properties, and whatever pathology considers a sufficiently negative consequence (in Schwartz's sense) for a condition to count as a pathological.

Objections to the etiological theory of pathological condition

Objection (i)

Given that the ability of the reproductive organs to aid reproduction is the explanation (or at least one of the explanations) for their existence, ETPC must account for the fact that the inability of young children to reproduce in not pathological.

Response (i)

We have not evolved to reproduce as young children.[16] Clause 2 specifies that functions are reference class relative, so any objection of this kind can be dismissed.

Objection (ii)

ETPC fails to distinguish between old age impairments, and normal impairments.

Response (ii)

Boorse suggests that, given that different reference classes provide different biostatistical norms, some conditions are diseases for the elderly but not for young adults. In many instances, my conception does not 'accommodate' this, but I consider it an advantage of the etiological theory of disease. Surely Alzheimer's disease is a disease for the elderly as well as for the young (even if most elderly people were to suffer from the condition, which is an increasing worry in our aging population); surely a pathologist can identify coronary heart disease in

DOI: 10.1057/9781137552921.0008

the elderly as much as she can in the young. The etiological conception of disease is grounded by evolutionary theory, so assuming our brains and hearts have not evolved to dysfunction, the pathological conditions of young adults and the elderly coincide. That is not to say these conditions should be classified as identical – the same impairment to a natural function might have to be classified and treated differently depending on the age of the patient. Early onset Alzheimer's and early onset Parkinson's, for example, are treated as different medical conditions to the Alzheimer's and Parkinson's diseases suffered by the elderly. Nonetheless, both are pathological conditions, and the target is only to differentiate between health and disease states – it is not a means of classifying diseases.[17]

Objection (iii)

One might reiterate Cooper's objection to Boorse's theory, that there are more reference classes than simply age, sex, and species-type (see *objection (ii)* in this chapter).

Response (iii)

Applying a Neander-style etiological account of function to a theory of disease also deals with the 'too many reference classes' objection. If one takes the natural function of a trait to be the reference class relative explanation for its existence, and a disease to be the malfunctioning of that trait, what is statistically typical for a reference class becomes irrelevant. It does not matter whether there is only one member of a reference class, nor does it matter how many reference classes there are.[18]

Objection (iv)

ETPC is not value-free. Etiological approaches to natural function have a normative component – they tell us what physiological subsystems are *supposed* to do, not just what they, in fact, do. Given that ETPC is grounded by an etiological account of natural function, it is inherently value-laden, and thus not a suitable candidate for the theoretical notion of disease.

Response (iv)

Given that the ETPC rests on an etiological account of function, the fact that the view has normative components is undeniable. However, Karen

DOI: 10.1057/9781137552921.0008

Neander (1991) embraces the normative and teleological qualities of her etiological account of natural function, asserting that 'the biological notion of "proper function" can be both teleological and scientifically respectable' (1991, 454). ETPC contains no qualities that are not 'scientifically respectable'; further, as I argued earlier in this chapter, that there is a normative component to a theoretical conception of disease, does not render it implausible, since it does not permit disease states to differ when physiological states are identical.

Conclusions

In this chapter I have detailed the most commonly discussed naturalist conception of disease, the BST. Many have objected to the BST, the most common objection being that the view is not, as it claims to be, value-free – but Boorse is right to say that this objection is misguided. For a concept to be value-free, it need only contain no value-laden ideas, and the BST does not. The BST is, however, susceptible to Guerrero's Cambridge change objection.

The FNC approach is immune to Cambridge changes. Schwartz introduces a component that takes into consideration the negative consequences of a condition, so as populations become healthier (or less healthy), the line which demarcates the healthy from the diseased portions of the population shifts accordingly. Schwartz's approach, if one is willing to accept a non-etiological conception of natural function, is a plausible theoretical conception of disease.

In the final section I proposed the etiological theory of pathological conditions. According to this view, diseases are impairments to natural functions with negative consequences (to a degree determined by pathology), where the natural function of a trait is that which explains its existence in the organism. This differs fundamentally from the etiological conception of disease in clinical medicine, since no two individuals could be in the same physiological state, but differ with respect to their disease status. The view is value-laden, but only in a perfectly scientifically respectable way – the only value-laden qualities of the thesis are grounded by evolutionary theory, and the informed choices pathologists make, concerning which impairments to biological function have sufficiently negative consequences, to be deemed pathological. It is therefore

DOI: 10.1057/9781137552921.0008

a viable alternative for those who prefer etiological accounts of function to their non-etiological counterparts.

Part I   Conclusions

Given that clinicians are concerned primarily with resolving the complaints of their patients, it is unsurprising that the conclusion reached in Chapter 1 involved a large normative component – if a condition is not bad for an individual, then it is not a disease *for her*; the very same condition, may, however, be a disease for someone to whom it is harmful.

The concept of 'disease' and the concept of 'harmful, medically treatable condition' differ significantly (this is true for both the clinician and the pathologist) – they must do, since many cases of the latter are not diseases. To accommodate this, a conceptual analysis of disease in clinical medicine must include a concept of proper function. I presented a number of options, but concluded that the most viable concept of disease for the clinician is one of 'harmful functioning'. This requires an etiological view of natural function; accounts for the clinician's need to make evaluative choices; and successfully distinguishes between diseases, and those harmful, treatable conditions that do not fall into this category (such as unfortunate tattoos).

In Chapter 2 I focussed primarily on Boorse's BST, but I ultimately endorse an etiological account – the ETPC. This account of pathological condition has a normative component, but the normative component does not concern the harm caused to the individual patient, in Cooper's (2002) sense; it is normative only because it is grounded by an etiological conception of natural function. Given that the presupposed etiological account of function is scientifically respectable, and that the view does not factor in whether or not a condition is harmful to the individual (in Cooper's sense), the ETPC being value-laden is unproblematic.

We can see that the clinician's concept of disease is not co-extensive with that used in pathology. For the clinician, if a condition is to be considered a disease, it must cause the individual patient harm; that is, it must be *bad* for the patient. For the pathologist, on the other hand, if a condition is a disease for one, it is a disease for all. The concept of disease, then, is context dependent.

DOI: 10.1057/9781137552921.0008

Notes

1 Unless stated otherwise, 'disease' and 'pathological condition' will be used interchangeably for the remainder of this chapter.

2 See objection (iii) to the BST.

3 Later revised to 'survival *or* reproduction', to accommodate reference classes that do not reproduce (Boorse, 2014)

4 Note that, according to Boorse, the concept of disease is not value-laden, so those who believe the biostatistical theory in some way degrades homosexuals misunderstand the thesis.

5 This is nicely consistent with contemporary thought in biology. See Dawkins 1976; Llyod 2007; Smith 1998

6 To have a pancreas producing subnormal quantities of insulin is to have one of a variety of forms of diabetes.

7 'Training', since marathon runners and cyclists etc. will often have slower heart rates than relatively sedentary people.

8 As stated, Boorse's hypothesis can accommodate 'race' as a reference class, anyway.

9 Where I have written 'pathology', Boorse writes 'medicine'. This replacement is useful to avoid confusion with other contexts in which the term 'disease' is used, and warranted, since Boorse explicitly states that he is analysing the concept of 'pathological condition'. I shall not replace 'medicine' with 'pathology' throughout this chapter, but the same substitution can usually be made.

10 It is worth noting that Boorse does not believe that those in the medical profession always have it right, so his thesis includes a small proviso: 'that scientists are sometimes confused, inconsistent, or (as with fever) empirically wrong about their subject' (Boorse, 2014, 713). The biostatistical theory, then, is not merely a descriptive thesis, but a prescriptive one – it tells us which physiological conditions *should* be deemed pathological.

11 For a detailed account of how constructivism, so understood, can dovetail with the BST, see Kingma 2012.

12 See Boorse (2002).

13 See Chapter 1 Section entitled 'The frequency and negative consequences approach, and line-drawing problem'.

14 It is worth noting that Kingma does not consider this a destructive objection, referring the reader to Hausman (2012). It is not within the scope of this chapter to investigate this further, however.

15 Or there is a limitation on functional ability caused by environmental agents.

16 In fact, the etiological theory of disease deems *fertile* children to have a pathological condition. The early onset of puberty, or 'precocious puberty',

DOI: 10.1057/9781137552921.0008

is a recognised pathological condition. Not only is fertility unsuitable for children because they are not sufficiently (mentally) mature, but precocious puberty causes the early maturation of bones, and consequently the early cessation of linear bone growth.

17   Of course, this argument flounders if physiological subsystems have evolved to dysfunction, but given the aims of medicine, medicine does not (nor should it) consider this possibility relevant to either clinical practice or pathology.

18   For the record, I am happy with Boorse's response that medicine (pathology) has chosen the reference classes, anyway.

DOI: 10.1057/9781137552921.0008

Part II
Disease and Causation

▶

DOI: 10.1057/9781137552921.0009

3
Concepts of Causation in the Philosophy of Disease

Abstract: *In this chapter I outline three conceptual analyses of causation in contemporary analytic philosophy, considering their actual and potential applications in both evidence-based practice and medical research. As with the concept of disease, I conclude that there is not one analysis best suited to medicine. The notion of 'what causation is' is context dependent, insofar as the pathologist, the clinician, and the epidemiologist must (and do) adopt different conceptions of causation.*

Keywords: Classification of diseases; counterfactuals; Disease and causation; dispositionalism; Mumford and Anjum

Smart, Benjamin. *Concepts and Causes in the Philosophy of Disease.* Basingstoke: Palgrave Macmillan, 2016. DOI: 10.1057/9781137552921.0010.

DOI: 10.1057/9781137552921.0010

Introduction

Causation is fundamentally important to all areas of medicine: the GP diagnoses the cause of the patient's discomfort, prescribing drugs, rest, and therapy, based on his or her knowledge of the determinants of disease, and of treatments (interventions); the pathologist sometimes looks to identify a cause of death; and the epidemiologist conducts studies to *discover* the determinants of disease, and both preventative and curative interventions.

In this and the next chapter I discuss two related, but distinct topics concerning causation in medicine. Chapter 4 looks at causal inference in public health. There I focus on the potential outcomes approach, a popular theoretical framework to effect measurement in epidemiology. In this chapter I outline three conceptual analyses of causation in contemporary analytic philosophy, considering their actual, and potential applications in both evidence-based practice (EBP), and medical research. As with the concept of disease, I conclude that there is not one analysis best suited to medicine. The notion of 'what causation is' is context dependent, insofar as the pathologist, the clinician, and the epidemiologist must (and do) adopt different conceptions of causation.

A complete discussion of the virtues and vices of the conceptual analyses dealt with in this chapter is a book in itself, so I do not set out to convince the reader to adopt any one of the views explicated. I do, however, try to present the most common objections, focusing in particular on those raised by Kerry et al in the context of EBP; where possible, I respond accordingly.

The following section looks at a reductive account of causation, according to which causation is a matter of counterfactual dependence (Mackie 1974; Lewis 1973a, 1973b, 2000; Hausman 1996). Crudely, the counterfactual thesis takes one event c to be the cause of another event e if and only if both c and e occur, and if c had not have occurred, then e would not have occurred. In this section I argue that the counterfactual account neither falls foul of the objections raised by Kerry et al, nor of a number of other objections that *may* have been raised by philosophers of medicine with similar motivations.[1] I ultimately argue that the counterfactual conception of causation is well-suited to the pathologist; however, although this reductive analysis might be 'what causation is' in the context of pathology, it is not useful for the clinician.

DOI: 10.1057/9781137552921.0010

Clinicians must consider the properties of individual patients before treatment, since the qualities of the patient are as relevant to the outcomes of interventions, as the properties of the interventions themselves. Clinicians often do not know how a patient will react to a stimulus (be it a potential determinant of disease, or a possible intervention); the best one can hope for, prior-to-the-fact, is (in general) an understanding of how the patient is likely to, or is 'disposed to' react. Following Eriksen et al (2013), and Kerry et al (2012), I first propose that clinicians employ a dispositional (or 'tendential' (Mumford and Anjum 2011)) conception of causation. I then claim that Mumford and Anjum's (2011) vector model of 'powers' is, for the most part, a far more accurate representation of causal reasoning in EBP, than that espoused by the counterfactual theorists.

To conclude this chapter I outline an unselective regularity account of causation, according to which causes are non-redundant conditions, together sufficient for their effects. I demonstrate that Rothman's (1976) sufficient-cause model to epidemiology is predicated on this analysis, and that the model is of no use to the clinician. However, I go on to propose a means of classifying diseases, using the regularity account of causation implicit in Rothman's model.

Causation as counterfactual dependence

In *An Enquiry Concerning Human Understanding* (1748) David Hume defined a cause to be an *'object, followed by another, and where all objects similar to the first are followed by objects similar to the second.* Or in other words, *where, if the first object had not been, the second never existed'* (Hume 1999).[2] Although this was not appreciated by Hume at the time, these are two very different approaches to causation. The first is often referred to as the 'naïve regularity account' (Armstrong 1983; Mumford 2004), a more sophisticated version of which will be presented (in the form of Rothman's sufficient-case model to epidemiology) later in this chapter; however, the second is a basic description of the still-popular counterfactual account of causation.

The counterfactual account states that 'If *c* and *e* are two actual events such that *e* would not have occurred without *c*, then *c* is a cause of *e*' (Lewis, 1973a, 563). I take this to be somewhat intuitive. Smoking caused Joe's lung cancer, since Joe smoked and got lung cancer, and if Joe had

DOI: 10.1057/9781137552921.0010

not smoked, Joe would not have got lung cancer. Drinking, on the other hand, did not cause Joe's lung cancer, since if Joe had not drunk, he *still* would have got lung cancer – with respect to Joe's cancer, smoking was a difference-maker (a cause); drinking was not.

Initial objections to the counterfactual conception

Before outlining the objections to the counterfactual analysis raised in Kerry et al's defence of a dispositionalist conception of disease, it is worth presenting two additional popular counterexamples to counterfactual theories of causation.

Objection (i)

Suppose a patient has a major stroke, and dies immediately. Given that she is not, in fact, simultaneously hit by a bus (or subject to any other life-threatening event), one can justifiably assume the stroke to be the cause of death. Nevertheless, on the face of it one can accept that if she had not had a stroke, she *might* have had a heart attack at that same moment, and died of the heart attack instead. If so, then one implicitly accepts that the counterfactual 'if she had not had a stroke, she would not have died' *might* be false. The counterfactual account, then, does not permit one to assert that the stroke certainly caused the patient's death (yet it clearly did).

To rule out precisely this kind of counterexample, David Lewis goes to great length to provide a methodology for evaluating counterfactuals. A detailed analysis of his view is well beyond the scope of this chapter, but crudely: Lewis suggests that there are real 'possible' worlds very similar to ours (the 'actual world'),[3] and that some are more like the actual world than others. He provides a list of criteria to test *how* like the actual world (that is, how 'close to' the actual world) these possible worlds are. The list includes a 'matching of local particular matters of fact' clause – again, crudely, if the world appears very similar to the actual world (in terms of what happens in it), it is likely to be closer to the actual world, than one that does not resemble it at all.

To evaluate the counterfactual, one considers the closest possible world in which (in this case) the patient does not have a stroke. This world probably looks exactly like the actual world until the moment the patient has the stroke. There is then a 'minor miracle' in the possible world, which permits the stroke *not* to occur. The possible world then runs on in accordance with the laws of the actual world, and we see whether, in

DOI: 10.1057/9781137552921.0010

this closest possible world in which the patient does not have a stroke, the patient dies of something else instead. If not, then the counterfactual is true, and one can assert that the patient died of the stroke.[4]

Lewis's approach to this kind of counterexample is very effective. However, talking of other possible worlds, miracles, and so on, is a little cumbersome in the context of this project. J. L. Mackie replies to the same kind of objection by introducing an 'in the circumstances' clause, to be expressed prior to the standard counterfactual; that is, he suggests one evaluates whether '*in the circumstances*[5] [e] would not have occurred if [c] had not' (1974, 31). Mackie's response is, in essence, very similar to Lewis's. Given the relatively simple way in which it is expressed, in the remainder of this chapter I speak mainly in terms of evaluating 'in the circumstances' counterfactuals. One should note, however, that when I do so, Lewis's possible worlds approach is presupposed.

Objection (ii)

The patient's death during an operation did not cause her to fail to survive, yet given that 'If the patient had not died during the operation, she would have survived it' is true, the conditional account of causation (as it is expressed here) implies a causal relationship.

Counterexamples of type (ii) occur only because the crude version of the conditional account outlined here does not assert that causes and effects must be, in Hume's words, 'distinct existences'. Mackie suggests 'it is sufficient to say that someone will not be willing to say that X caused Y unless *he* regards X and Y as distinct existences' (1974, 32). Similarly, Hausman expresses Lewis's conception as follows:

> L (*Lewis's Theory*) c causes e if and only if c and e are *distinct events*[6] and if c were not to occur, then e would not occur either. (Hausman, 1996, 55)

The patient's death and the patient's failure to survive are not distinct existences, since to not survive just *is* to die. This second counterexample, like the first, is not overly problematic for the counterfactual theorist.[7]

Kerry et al's objections to the conditional analysis in evidence-based practice

Kerry et al (2012) defend the view that EBP should assume a dispositionalist conception of causation – a conclusion that I later endorse. In the process, however, they criticise the counterfactual conception of causation, when applied to healthcare more generally. In this section I critique the objections they raise, as well as some related, potential objections that they do not. Although prima facie problematic, none of the objections are convincing (permitting the counterfactual view to be applied elsewhere in medicine).

Objection (iii)

In the next chapter I discuss the POA, a counterfactual method of effect-measurement employed in public health studies. It compares the actual outcome of a study with some contrary-to-fact supposition. The counterfactual claims the POA permit, however, must always specify the relevant property/properties of the comparison class; that is, exposure E causes outcome O *versus* some contrast class Z (smoking causes cancer relative to not smoking). In light of this, Kerry et al provide the following objection to counterfactual accounts of causation in EBP:

> In two groups, A (the intervention) and B (the control), there will be a certain proportion who achieve the outcome of interest, say 58% in group A and 42% in group B. Depending on the research question, power of the study, etc, statistical analysis will be performed to determine the significance of this difference. If significant difference is established, then the recommendation would be that, thus far, A is the preferred intervention compared with B. In other words, there will be a greater causal effect from A than B. But what does this say about the 42% of subjects who responded just as well with B? Again, the issue is that it looks like something causal did in fact happen to 42% of subjects in group B, but this cannot be accounted for in the way health science considers causation. (Kerry et al, 2012, 1009)

The point Kerry et al make here needs clarification.[8] Suppose I conduct a study, in which I compare a group of smokers who smoke, on average, 20 cigarettes a day (exposure A), with a group of smokers who smoke, on average, 10 cigarettes a day (exposure B). The study shows that smoking 20 cigarettes a day causes cancer versus smoking 10 cigarettes a day (the risk is higher if you smoke more). If one already knows that smoking 10 cigarettes a day increases risk of cancer relative to not smoking at all, a study showing that smoking 20 a day causes cancer versus smoking 10 cigarettes a day is interesting, since it provides evidence of what Hill (1965) calls a 'biological gradient'[9] (the fact that more you smoke, the worse the outcome, provides evidence for the general causal claim that 'smoking causes cancer'), but that particular

DOI: 10.1057/9781137552921.0010

study tells us nothing about the causal effect of smoking 10 cigarettes a day versus non-smoking.

Kerry et al are quite right about this. So what is their objection? There are two possible interpretations. The conclusion of the first interpretation involves making a category error; the conclusion of the second interpretation (that the study expressed does not tell us the causal effect of B *simpliciter*) is entirely unproblematic.

Interpretation 1: the category error

Kerry et al, at the end of their critique of the counterfactual approach, write: 'it is clear ... that the counterfactual conditional fails to get to the essence of what causation is.'

Health studies of the form described above are designed to measure the causal effects of exposures. They do admit of causal conclusions, but unlike Lewis's counterfactual theory, these studies have nothing to do with what causation *is*. If the objection is that this 'counterfactual' approach fails to tell us what causation is, then a category error has been made.

Interpretation 2: we learn nothing about B

Suppose one wanted to discover the effects of smoking 20 cigarettes a day *simpliciter* (strictly speaking, this makes little sense given the structure of epidemiological studies, but more on that soon) – one would not attempt to discover these effects by comparing a group who smoke 20 cigarettes a day, with a group who smoke 10 cigarettes a day: one would compare the 20 a day smokers with non-smokers.

The fact that some people in the study Kerry et al outline *responded* just as well with exposure B, as those who *responded* well with exposure A, implies that both group A and group B are being treated in some way; but when, in ordinary discourse, one asks 'what are the causal effects of B?' (that is, when one wishes to determine whether or not B is a cause of an outcome, and how much of an effect it has), strictly speaking (in epidemiological terms) one is asking 'what is the effect of B relative to no intervention'. A well-designed study of the same format can usually accommodate this causal question.

In short, *of course* measuring A relative to B will not tell us anything (or at least, not much) about the causal effect of B – but that is not the causal question asked by this health study. The question asked by the epidemiologist,

DOI: 10.1057/9781137552921.0010

here, is 'how effective is A relative to B?'. One *can* find out the causal effect of B relative to no intervention, but that is a different study.

None of this, however, has any bearing on the success of Lewis's reductive account of causation as counterfactual dependence, since the health study's 'counterfactual' approach to effect-measurement *is not the same thing*, as the counterfactual account of what causation that they set out to defeat.

Objection (iv)

Kerry et al state that 'Counterfactual theories...take causes to be the same as necessary conditions. This would mean however that birth is a cause of death, and having a back is a cause of lower back pain, that is the counterfactual condition in each of these examples demonstrates that if you had not been born, you would not have died; if you did not have a back, you would not have lower back pain' (2012, 1009).

Some might claim that this is, even at face value, unproblematic for the counterfactual account, since having a back *is* a cause of back pain. Most, if not all events, are causally affected by millions of others. Suppose Sam is playing golf, and holes a 20 foot putt. She strikes the ball sweetly with her putter in the planned direction, at a pace fast enough to reach the hole, but not so fast that it jumps over – Sam's putting stroke is a cause of her holing the putt, but it is not the only causally relevant factor. She might wish to focus on the good putting stroke, and assert that her stroke was *the* cause of the ball dropping into the hole, but the wind, the grain of the grass, the shape of the ball, the slope of the green, small indentations on the putting surface from footprints and pitch-marks, the force of gravity, and many, many other variables feature in determining where the ball finishes. Were any of these variables different by a significant degree, the ball would not drop. Lewis states:

> We sometimes single out one among all the causes of some event and call it 'the' cause, as if there were no others. Or we single out a few as the 'causes', calling the rest mere 'causal factors' or 'causal conditions'. Or we speak of the 'decisive' or 'real' or 'principal' cause. We may select the abnormal or extraordinary causes, or those under human control, or those we deem good or bad, or just those we want to talk about. I have nothing to say about these principles of invidious discrimination. I am concerned with the prior question of what it is to be one of the causes (unselectively speaking). My analysis is meant to capture a broad and nondiscriminatory concept of causation. (1973a, 558–559)

DOI: 10.1057/9781137552921.0010

For those, like Lewis, who refuse *selective* theories of causation, (iv) looks to be unproblematic. But medicine is surely not at liberty to adopt such a conception – not many doctors would consider the *presence* of the aorta to be a cause of heart disease!

Fortunately, there are selective accounts of causation on offer (Menzies 2004; Schaffer 2005; Broadbent 2008). Again, a long discussion of causal selection is beyond the scope of this chapter – but let us briefly consider two options: Mackie explains that:

> causal statements are commonly made in some context, against a background which includes the assumption of some *causal field*. A causal statement will be the answer to a causal question, and the question 'What caused this explosion?' can be expanded into 'What made the difference between those times, or those cases, within a certain range, in which no such explosion occurred, and this case in which an explosion did occur?' Both cause and effect are seen as differences within a field; anything that is part of the assumed (but commonly un-stated) description of the field will, then, be automatically ruled out as a candidate for the role of cause. (1974, 34–35)

Mackie thus agrees with Lewis insofar as he takes causes to be difference-makers, but disagrees insofar as he does not adopt his 'broad and non-discriminatory' approach. Causal questions presuppose a causal field[10]; that is, a set of circumstances, no part of which can be deemed a cause. When one asks 'what caused Bob's lung cancer?' the existence of his lungs, his being alive, and so on, all form a part of the field, and so *ex hypothesi* cannot be causes. Bob's smoking, on the other hand, does not form part of the causal field, so if the conditional states that if he had not smoked, he would not have had lung cancer, one can identify smoking as a cause of his cancer.

Jonathan Schaffer defends 'causal contextualism' (versions of which are also defended by Woodward 2003; Maslen 2004), the thesis that: 'A single causal claim can bear different truth values relative to different contexts, where this difference is traceable to the occurrence of "causes", and concerns a distinctively causal factor' (Schaffer, 2012, 37). One might ask, for example:

1. Why did *these* elephants cross the river?

The emphasis on 'these' implies that one is questioning the cause of *this specific herd* of elephants crossing the river, rather than the herd next to them. Perhaps the answer is: 'because there wasn't enough food for the two herds, and this was the only herd *able* to cross the river'. On the other hand, one might ask:

2. Why did this these elephants *cross the river*?

Here, the emphasis suggests one is questioning why the herd of elephants crossed the river rather than, say, walking elsewhere. Perhaps the answer is 'because the only area with food nearby was the other side of the river', but it is not 'because there was not enough food for the two herds of elephants.'

For Schaffer, causal claims are sensitive to contrasts, whereby 'what shifts with context are the contrasts in play, where contrasts are specific possible alternatives to actual events' (45). With scenario 1, the contrast classes involve the other herd of elephants attempting to cross the river; with scenario 2, the contrast classes involve the herd of elephants that crossed the river, walking upstream, staying where they are, moving inland, and so on.

The contrastive view, he writes, can 'be plugged into a simple counterfactual [conditional] test for causation by replacing the supposition of the non-occurrence of c or e [cause or effect], with the supposition of the occurrence of the associated contrast [c^* or e^*]. So, for instance ... one might hold that c rather than c^* causes e rather than e^* iff (roughly) if c^* had occurred then e^* would have occurred' (45–46).

Kerry et al claim that the counterfactual theory implies that having a back is a cause of back pain – but if Schaffer's view is coherent, the counterexample can be defeated, since no contrast class for *any* reading of the question 'why does this man have back pain' involves a man with no back. I do not pretend to have provided a detailed critique of either Mackie's or Schaffer's conception. I outline the views here only to show that theories of causal selection exist, and that they are not prima facie implausible.

Objection (v)

Suppose that, in the same patient, at the same time, two events (say, A and B) occur that are individually sufficient for the same disease D. Neither is a necessary condition, since neither the counterfactual 'if A had not occurred, D would not have occurred', nor the counterfactual 'if B had not occurred, D would not have occurred' is true. Therefore, given that causes are necessary conditions, neither A nor B is a cause of D.

This is a common occurrence – effects are often causally overdetermined. Suppose a firing squad of four shoot a soldier. It is not

DOI: 10.1057/9781137552921.0010

true of any member of the firing squad that, if he had not pulled the trigger, the soldier would still be alive. It would clearly be very damaging to the counterfactual account, if it implied that none of the bullets killed our soldier – he is dead, after all. One can, however, deny that any one of the bullets individually caused the soldier's death, and suppose that all the bullets, together, caused the soldier's death (for it is true that if none of the bullets hit the soldier, then the soldier would not have died).

Rather than considering either A or B causes individually, one must take the cause of the disease to be A and B together. The counterfactual 'if neither A nor B occurred, then D would not have occurred' is true, so the problem disappears.

Objection (vi)

'If [a randomised controlled trial (RCT)] failed to show a significant difference between two intervention groups, but in both groups a treatment effect was observed, then the counterfactual stance would have to support the statement that neither intervention caused the effect.' (Schaffer, 2012, 1009). This issue, I believe, is either illusory, or can be resolved with Mackie's 'in the circumstances' clause.

First, a brief explanation of a basic RCT: a standard RCT involves two maximally causally homogenous experimental groups (crudely, the groups are causally homogenous insofar as they would produce the same outcomes, from the same exposures, for the same reasons, were neither group subjected to the exposure the RCT is investigating), from a subpopulation of the target population (the population to which the results are to be 'exported' – perhaps, but not necessarily, the general population (Cartwright 2007)). One of the experimental groups is then exposed to some humanly manipulable intervention (e.g. a drug, a particular diet, etc.). After a specified period of time, the two groups are compared, and given that they were otherwise causally homogenous, any significant differences are the effect of the exposure. In terms of counterfactuals: suppose there are two experimental groups in an RCT, group A and group B. Group B is exposed to E and group A is not. If the risk (statistical frequency),[11] R_1, of some outcome O is higher in group B at the end of the study, than the risk, R_2, of O in group A, then one establishes a causal connection between E and O, since if group B had not been exposed to E, then R_1 would be equal to R_2. E was a difference-maker, and hence a cause.

DOI: 10.1057/9781137552921.0010

Unlike the simple case above, some RCTs involve two or more intervention groups. In a case where there is no control group, one can only compare the results of the two intervention groups. Of course, without reference to *any* other data, this tells you very little. After all, when the outcomes are the same, the two intervention groups either show the same causal effect, or no effect whatsoever. In this type of trial, just as with objection (iii), the question asked by those conducting the studies is merely a comparative one concerning effect-measurement, and hence to use this as a criticism of the counterfactual analysis of causation is entirely misguided.

Sometimes there might be two interventions *and* a control group. In these cases, there are *in effect* two distinct RCTs (let us call these RCT1 and RCT2) being conducted. Each RCT has an exposure group (X1 and X2 respectively), and they share an otherwise causally homogenous control group (Y). RCT1 tests exposure A in X1, and RCT2 tests exposure B in X2. A and B are distinct, yet produce the same causal effects in X1 and X2 respectively (that is, the differences in outcome between X1 and Y, and between X2 and Y, respectively, are identical). One might claim that because the causal effects of A and B are identical, neither can be deemed a cause, but again, this would be misguided. If the trial has been well designed, then when one adds the 'in the circumstances' clause (which fixes the control group Y) the right counterfactuals come out true: '*in the circumstances*, if (contrary to fact) group X1 had not been exposed to A, then the outcomes of X1 and Y would *not* have differed' – A is thus deemed a cause according to the conditional account, since it was a difference-maker.

Conclusions

Kerry et al's conclusion is that it 'is clear ... that the counterfactual conditional fails to get across the essence of *what causation is*[12]' (2012, 1009). It is far from clear, though, that any of the arguments presented in this chapter establishes this. Lewis's conception of causation as counterfactual dependence is, at least on the face of it, a fairly robust reductive account of 'what causation is'. Furthermore, the view is very well suited to the pathologist, whose job is very often to find and monitor the cause of a patient's illness, or in some instances, the cause of death (that is, to find C in the counterfactual: Were the patient not to have had condition C, she would not have died). Nonetheless, the counterfactual approach is not well suited to all medical contexts.

DOI: 10.1057/9781137552921.0010

In Chapter 4 I outline in more detail how the epidemiologist discovers the causal effects of interventions, focusing on the POA. We shall see that the methodology of this theoretical framework, like RCTs, draws heavily on the counterfactual account of causation – but the experimental justification of general causal claims, although (of course) fundamentally important to EBP, does not reflect the day to day activities and thought-processes of the clinician. Clinical medicine focuses on the signs and symptoms of individual patients, and how best to treat the inferred conditions. Answering the *clinician's* questions requires a new conceptual framework. In the next section I outline a dispositional conception of causation, and show how modelling causation in terms of dispositions is more useful to EBP than the counterfactual conditional approach.

Clinical medicine and the dispositional account of causation

Just as salt is disposed to dissolve in water, and fragile objects are disposed to shatter, humans are disposed to contract diseases. Many of our physiological subsystems are disposed to dysfunction, and these dispositions can be triggered by the dispositions of invading disease entities (human papilloma virus is disposed to cause genital warts, for example), by those of our internal organs, by those of chemical substances (poisons), and so on. Epileptics are disposed to have seizures after excessive alcohol consumption; that is, alcohol is a trigger, or 'stimulus condition' for seizures, which 'manifests' when consumed by epileptics (in the absence of interfering factors). Some are disposed to go into anaphylactic shock when stung by bees, and smoking can trigger one's disposition to get lung cancer.

The dispositional conception is a non-reductive account of causation (Mumford and Anjum 2011) that in this context, takes individuals to 'tend towards' disease, and the determinants of disease to 'tend towards' causing disease (Eriksen et al 2013; Kerry et al 2012). In all the above cases, the stimulus conditions – viruses, poisons, alcohol, bee stings, smoking, sugar-ingestion, and so on – interact with the dispositions of patients, and *can,* but do not necessarily, lead to changes in the functional efficiency of physiological subsystems. In short, (i) every physiological change that takes place while diseased (indeed, any physiological process at all!) seems to be explicable using dispositional terminology; and

DOI: 10.1057/9781137552921.0010

(ii) the justification for clinical decisions seems to be, at least partly, grounded by what the clinicians see to be potential risks and benefits of the prescribed interventions; which, again, are explicable in terms of dispositions.

Unlike the counterfactual account, this view seems well-suited to clinical practice. Humans are disposed to contract malaria when bitten by certain malaria carrying mosquitos. Clearly, in the circumstances considered 'normal' in non-malarial zones, this is not a problem, since no malaria carrying mosquitos are around to trigger that disposition – but of course that can change when one enters Kruger National Park. Jane (like most other people) is disposed to contract malaria if bitten by a malaria carrying mosquito,[13] and given the presence of these mosquitos in Kruger National Park (depending on the time of year and where one is in the park), she might contract the disease when she visits. The doctor cannot remove the disposition of mosquitos to spread malaria, but by prescribing the prophylactics, she can remove Jane's disposition to contract malaria prior to her entering the park.

The Mumford-Anjum model

In some regions, an important aspect of clinical medicine is the treatment of poisons using antidotes; again, this can easily be accommodated by a dispositional conception of causation. Suppose Rachel is bitten by a snake and taken to hospital for treatment. Using the vector model of 'powers' proposed by Mumford and Anjum (2011, 39),[14] Figure 3.1 represents the reaction Rachel's physiological response to snake venom after an antidote has been injected. The vectors represent causes/conditions that dispose Rachel either towards health, or towards disease. The length of the arrow determines strength of that contribution, and the direction of the arrow determines whether the condition disposes towards health, or towards disease. Take a to represent the disposition of the brain to bleed when subjected to snake venom, b to represent the disposition of her intestines to be damaged by snake venom, and c to represent the dispositions of the snake venom when injected into the blood. These causal factors (dispositions) all contribute towards the dysfunction of Rachel's physiological subsystems, that is, towards disease. Now take d to represent the pre-existing dispositions of Rachel's physiological subsystems to 'tend towards' health, and e to be the dispositions of the antivenin. R represents the resultant vector, which indicates whether,

DOI: 10.1057/9781137552921.0010

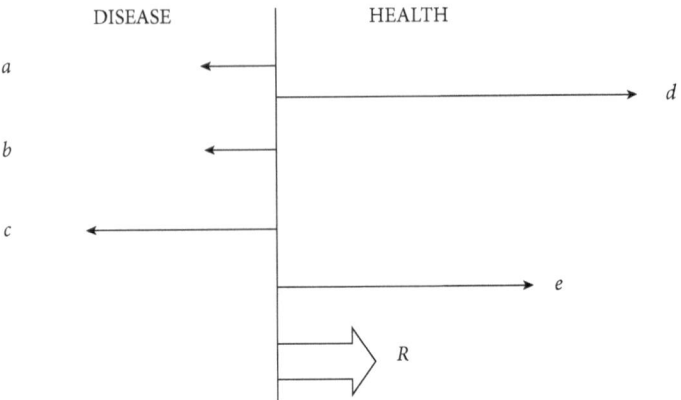

FIGURE 3.1   *The Mumford-Anjum model*

given all the causally relevant states of affairs, Rachel is disposed towards health, or towards disease:

Given the injection of the antivenin, R shows that Rachel is disposed to recover, since the sum of the vectors that contribute towards health, outweigh the sum of the vectors that contribute towards disease. If she had not been injected with the antivenin *e*, however, the sum of *a*, *b*, and *c* (that is, the dispositions of her body to be affected by snake venom, and the dispositions of the snake venom itself) outweigh *d* (the dispositions of Rachel's subsystems, under normal circumstances, to maintain a healthy state), and so Rachel would have been disposed towards disease. The strength of the disposition towards disease or health is thus represented by the size of the resultant vector.

If one has a 'chronic disease course' or a 'progressive disease course'[15] (Scheuermann et al, 2009), then the task of the clinician is to make the resultant vector (which, if it has magnitude, must point towards disease) as small as possible; and, if the disease can be cured, the task of the clinician is to make sure the resultant vector points towards health, with as great a magnitude as possible – to tend towards a quick recovery. Note that the model can represent any two properties. One might, for example, be interested in reducing pain, in which case, the vectors modelled would point not towards health and disease, but towards pain and comfort. Of course, just because R tends towards comfort, or towards health, that does not guarantee comfort or health – but this is a virtue of the thesis. A dispositionalist reading of cause allows one to assert the

DOI: 10.1057/9781137552921.0010

(intuitively correct) general causal claim that 'arsenic causes death', since the dispositional quality of that claim (that is, the fact that death is not *necessitated* by arsenic) is implicit in the assertion.

The Mumford-Anjum model, when applied to clinical medicine, is concerned with the dispositions of individual patients, and the effects of interventions on those individuals (as opposed to the effects of interventions on populations); the model can represent whatever qualities the clinician and patient are interested in; and the model accommodates the fact that clinicians regularly do not *know* how their patient will react to interventions during a disease course – they must operate on the basis that the right interventions dispose the patient towards health. The vector model, then, looks to be an excellent representation of the clinician's concept of causation.

The classification of diseases, and the sufficient-cause model of causation

In this section I discuss the sufficient-cause model of causation, and the related regularity model proposed by Mill and developed by Mackie.[16] This model does not represent the practices of the clinician, since it is not a tendential account. Nevertheless, that the sufficient-cause model does not reflect clinical practice does not make it misrepresentative in the stronger, metaphysical sense; that is, the tendential concept of causation employed by clinicians is not deterministic, but that does not imply that *as a matter of fact*, determinism (in medicine, and more generally) is false. It may be the case that clinicians must employ a tendential model, but nonetheless true that certain sets of conditions do necessitate disease states. Given that medicine is a pragmatic discipline, one might think this metaphysical claim somewhat irrelevant. However, outlining this view is not a fruitless task, since the sufficient-cause model is central to the method of disease-classification I later propose.

Consider Jane, a Kruger National Park safari guide – Jane is a little laissez faire about prophylactics, but when she feels there is a high risk of malaria, she takes them nonetheless. When deciding whether to take her pills, she considers the following factors:

1   *When* she is working. In mid-winter, there are very few malaria carrying mosquitos anywhere in Kruger National Park.

DOI: 10.1057/9781137552921.0010

2   *Where* she will be based – when there are mosquitos in the park, they are restricted to the northern region.

3   The quality of the insect repellent – even if there are lots of malaria carrying mosquitos around, she will not be bitten, given an effective insect repellent. With this in mind, she creates the following model:

Jane believes (quite reasonably) that diseases are not usually determined by a single cause. She also believes that if criteria (i) to (iii) obtain, and she does not take prophylactics, she *will* get malaria – the conditions represented by Figure 3.2, then, are together sufficient for contracting the disease. Furthermore, Jane's model represents what she deems to be the *only* jointly sufficient conditions for malaria (which in this case, rightly or wrongly, are translatable to what she deems to be the jointly sufficient conditions for being bitten by a mosquito).

If conditions (i) to (iii) are met, then Jane will take prophylactics, since if she does not take the sufficient-cause of malaria obtains – in taking the prophylactics, she removes a segment of the pie-chart, and so the sufficient-cause does not obtain, and she does not contract malaria. Jane is using Rothman's 'sufficient-cause model to epidemiology' (Rothman 1976; Rothman et al, 2008, 8) (of course, there may be many pie-charts for any one disease – there is more than one way to catch hypothermia – but every disease has at least one such chart).

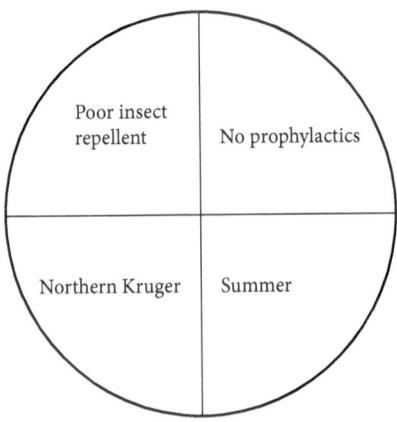

FIGURE 3.2   *Malaria's 'sufficient-cause'*

DOI: 10.1057/9781137552921.0010

As I have already stated, any suggestion that Rothman's model accurately represents clinical practice would be misguided. There are two main reasons for this.

1 Either one deems *all* the conditions mapped in the pie-charts to be causes, or one takes the cause of disease to be the *entire* sufficient-cause. Neither, of course, is compatible with clinical practice.

Rothman's model is unselective, so (once again) the 'presence' of lungs becomes a cause of lung disease. In these cases, however, one cannot appeal to either the causal field or contrast class responses. Causal fields and contrast classes are fixed by the nature of the causal question, and Rothman's sufficient-cause model does not ask causal questions; it merely provides information concerning jointly sufficient conditions for contracting token diseases.

2 The model is inherently deterministic.

As detailed in the previous section, clinicians cannot operate in this way. Clinical medicine, at least on the face of it, presupposes a dispositional conception of causation.

Rothman's model, despite not resembling the clinician's concept of cause, is nonetheless important for both clinical medicine and epidemiology. In the next chapter I discuss theoretical approaches to causal inference and effect measurement in epidemiology, concluding that Rothman's model is essential if one is to capture nonmanipulable causes in one's epidemiological framework. Here I show how the model (although expressed in a slightly different way) can ground an effective means of classifying diseases.

On the classification of diseases[17]

Identifying the necessary and sufficient conditions for individuating and classifying diseases is a matter of great importance in the fields of law, ethics, epidemiology, and of course medicine. Here I propose a 'causal classification of disease' (CCD). Note that this is not a metaphysical account of what a disease *is*, either from an ontological or conceptual perspective, but an account of how to individuate one disease from another (the cause of a disease is not the disease itself!).

When a patient becomes ill with influenza F one might determine the contraction of the influenza virus V to be the cause. However the virus

DOI: 10.1057/9781137552921.0010

is not the only factor involved in the patient contracting the disease. She has had a particularly poor immune system since she developed AIDS, P. This contributed to her contracting F, as *ceteris paribus* without P her immune system would have been strong enough for her to fight off the infection. Neither V nor P is individually sufficient for F, but both are causally relevant and jointly sufficient. Despite their being individually insufficient, we might nonetheless consider the cause to be F, or P, *or* both, depending on what information we consider most salient.

Furthermore, what appears to be the 'salient' cause often acts as an etiological agent for more than one disease (sometimes, in the same patient, at different times). Consider a case where the HPV virus causes a patient to manifest genital warts as a young woman, and at a later date the virus is deemed to be the cause of her cervical cancer (see figure 3.3). Although the two distinct diseases are caused by the same etiological agent, one can see that the invading organism does not tell the whole causal story.

According to 'multifactorial accounts' (Broadbent 2009), every token disease has numerous causes that are only jointly sufficient for it.[18] Furthermore, there are likely to be numerous distinct jointly sufficient conjunctions of causal conditions. This account is thus captured by Rothman's sufficient-cause model. That model itself draws heavily

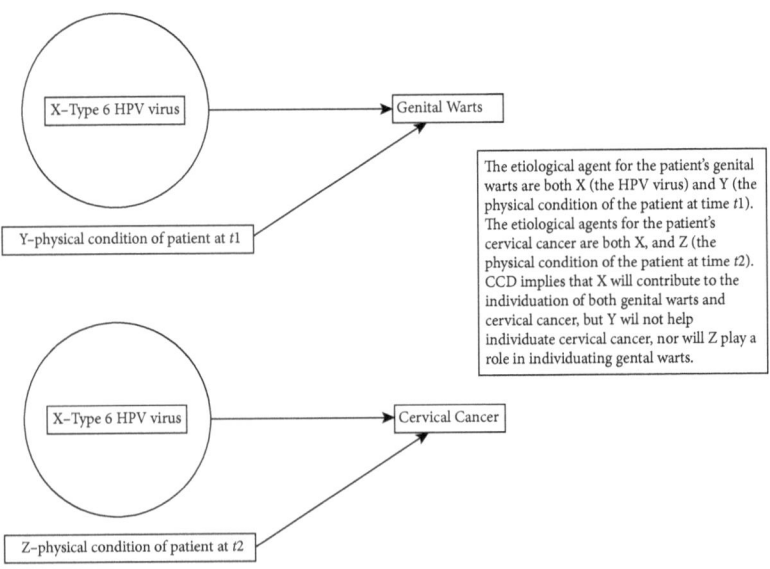

The etiological agent for the patient's genital warts are both X (the HPV virus) and Y (the physical condition of the patient at time *t*1). The etiological agents for the patient's cervical cancer are both X, and Z (the physical condition of the patient at time *t*2). CCD implies that X will contribute to the individuation of both genital warts and cervical cancer, but Y wil not help individuate cervical cancer, nor will Z play a role in individuating gental warts.

FIGURE 3.3   *Multifactorial accounts of disease*

DOI: 10.1057/9781137552921.0010

on Mackie's 1973 'regularity account of causation', which I use here to provide a means of individuating diseases.

When we consider what the causes of cardiovascular disease are (on a more general scale), we have a disjunction of conjunctions of causal conditions. Cardio vascular disease (CVD), for example, can result from 'XY or QY or...' Where '...' represents a finite number of conjunctive conditions; we may assert that CVD is always preceded by 'XY or QY or...' Conversely that all 'XY or QY or...' are followed by CVD. It is thus a regularity account in Hume's sense, insofar as a cause is constantly conjoined with its effect.

A disease obtaining might also be partially reliant on the *absence* of certain occurrences. In the CVD case, for example, lots of exercise might 'mask'[19] the disease. These are welcome additions to the causal classification model, as many diseases, such as scurvy and rickets, are the result of a *lack* of something (in these cases, vitamin C and vitamin D respectively). The result picture looks like this: 'Z iff XYnot-D or QYnot-D or...'[20] This model is clearly unselective, but in the context of disease-classification, this is a virtue, not a problem.

Given the removal of 'selection' from the process of cause-identification, the exhaustive list of conditions to be satisfied for most effects (for our purposes these are diseases, but this is a general account of causation) is enormous. Mill writes, though, that contrary to our every-day conception 'the cause...philosophically speaking, is the sum total of the conditions positive and negative.'[21] Following Mackie, let us call this 'philosophically speaking' cause, the 'full cause'.

According to CCD, diseases are to be individuated by their full cause; that is, the disjunction of conjunctions of events jointly sufficient for contracting the disease.

Pragmatically speaking, the 'full cause' is clearly not what we refer to in standard causal talk, either within or outside of an epidemiological/medical context. When we identify causes in a medical context the salient information is that which leads to diagnosis, prognosis, explanation, and the identification of suitable treatments for the disease. Indeed, a medic need not even identify *precisely* what disease a patient is suffering from in order to (non-accidentally) prescribe the correct treatment. Many antibiotics, for example, will successfully treat a large number of different bacterial infections, so in many cases to help a patient the medic need only identify the infection as bacterial. The salient cause(s) (where cause is being used in the every-day sense of the term; that is, one of the conditions in one of the conjuncts that partly comprise our full cause) and corresponding treatment(s) are thus often associated with many different diseases.

Given that one etiological agent can apply to many different diseases, using a single, salient causal factor (*inus* condition – see footnote 20) is not sufficient for a CCD. That said, once we take the cause to be the cause *qua* 'cause philosophically speaking', no two diseases will have the same full cause.

Objections to the causal classification model

Objection (i)

Mackie's general account of causation implicitly assumes that our conceptual framework is deterministic, but it is not obvious that a conceptual account of disease individuation can be so. Epidemiology is rife with talk of chance: 'chance of infection', 'chance of survival', 'chance of going into remission', and so on.

Response: Chance, in medical contexts, tends to refer to risk, whereby one's assigned chance of survival/recovery is determined by the percentage survival rate of similar patients with similar ailments, at similar stages of the disease, in similar environmental conditions, and so on. Frequentist probabilities do not rule out deterministic conceptions of causation in medicine – 'smoking raises the (statistical) probability of getting lung cancer' and 'lung cancer has its identity fixed by a (deterministic) full cause' are perfectly compatible claims.[22]

Objection (ii)

Different stages of a disease have different 'full causes', so CCD would identify them as different diseases.

Response: The model identifies diseases by the causal conditions of their initial contraction. Any stage of a disease could be traced back to the conjunction of events that caused the first stage of the disease – CCD individuates a disease by the full cause only of its first stage (that is not to say that a later stage of a disease cannot be individuated by the full cause of a disease-*stage*, but this is a virtue of the model).

Objection (iii)

Assuming determinism, and the transitivity of causation, the cause of any disease in person *P* is also the cause of any subsequent disease

P might contract. Now suppose P has infectious disease D, which weakens her immune system, and as a result she contracts disease D^*. For practical purposes it is (let us suppose) useful to treat D and D^* as distinct diseases, but how can CCD accommodate this, when it implies that D and D^* can be seen as different stages of the same disease?

Response: The aim here is to find a means of individuating diseases, that is, for no two diseases to satisfy the same conditions – CCD succeeds in this regard. CCD does not tell us which full causes identify diseases and which do not, but if D and D^* are distinct diseases (which we have agreed they are), then they each have their own unique full cause. Although it raises an important issue, objection *iii* does not defeat CCD.

Objection (iv)

'[I]f all we say about diseases is that they have many and diverse causes, it is not clear what more we can hope for than a catalogue. And the catalogue will never be complete, because in practice the causes of even the most ordinary event are beyond human counting' (Broadbent, 2009, 308). Here Broadbent makes the point that the full cause (or something like it) cannot be used in general explanation, and that since diseases are 'kinds of ill health for which general explanations are available', a catalogue of causes is an inadequate means of classifying disease.

Response: I agree with the motivations for Broadbent's objection. We classify diseases for practical purposes, primarily in the hope that we can better cure and/or prevent ill-health. To do this we first need the model to provide general explanations for ill-health, but does CCD fail in this regard? Broadbent claims that a model of disease *purely* comprised of lists of causes does not explain why we differentiate between different diseases, 'it says nothing about what disease is, or about why we distinguish between diseases in the way that we do.'[23] However in token cases (and here I speak of causal sequences in general, not just those concerning disease), causal explanations of particular events can be very useful. With token cases of ill-health, causal explanations can undeniably help us determine which procedures should be followed to restore good health. So why not generalise this claim, that is, from the cause of a disease-token to the full cause of a disease-type? If we identify a disease by its causes, we explain the ill-health – diagnosis becomes the identification of causal explanation (more specifically, the identification of the range of possible causal explanations), and points towards possible treatment(s)

DOI: 10.1057/9781137552921.0010

for the disease-type diagnosed. Furthermore, Broadbent's objection is raised against *bare* multifactorial accounts – I do not propose that all there is to a disease is a list of causes; rather, I take diseases to be the failure of physiological subsystems to perform their natural functions, and these diseases can be individuated by lists of causes.

With respect to the second of Broadbent's worries (that 'the causes of even the most ordinary event are beyond human counting'), it is indeed pragmatically impossible to know *all* the causal conditions and their ordering for any one disease, but the practical inability to fully pick out all the identity conditions for a disease does not significantly jeopardise the position. The worry is that diseases become 'unknowable noumena', but this is not an uncommon phenomenon in the sciences. Brian Ellis's scientific essentialism (2001, 2009)[24] takes the aim of science to be to discover the essential properties of natural kinds, but not knowing all the essential properties of a particular natural kind does not entail that the natural kinds are unknowable, nor does it make scientific essentialism wholly unappealing. Just as discovering unknown essential properties of natural kinds refines our scientific knowledge and understanding of the natural world, discovering further possible causes of a disease improves our knowledge, understanding, and ability to manipulate the disease. We may have to concede that knowing an individual disease's full cause/ identity conditions is beyond the scope of human understanding, but this does not make the position implausible.

Conclusions

In this chapter I have considered three conceptions of causation: causation as counterfactual dependence; dispositionalism; and a sophisticated form of the regularity conception of causation. Each seems to be appropriate for some areas of medicine, but not for others.

When one questions whether one event caused another, Lewis's counterfactual conception seems to capture the issues at stake. Did the proposed cause actually happen, and would the effect occurred had that proposed cause not? This conception, it seems to me, is very likely to be that employed by pathologists when deciding cause of death, and perhaps in the diagnostic stages of clinical practice. Furthermore, the objections to the counterfactual theory raised in this chapter are all ultimately unconvincing, so there is little reason to remain sceptical of the view in the contexts identified. However, the counterfactual approach is ill-equipped to deal with clinical practice.

The clinician is in the business of explanation and prediction; she must explain why a patient is feeling ill, and having explained the illness, predict which interventions will return her patient to a state of good health, in the fastest possible time. The counterfactual account of causation may be sufficient for the explanatory aspect of the clinician's role, since this is to provide 'after-the-fact' causal explanations. However the predictive side cannot be dealt with in this way. From a conceptual perspective at, least, interventions do not seem to necessitate outcomes. Furthermore, patients with the same disease will very often react differently to the same medication, since the outcome of an intervention depends as much on the properties of the patient as it does on the properties of the medication (penicillin will cure Jim's chest infection, but kill John, for John is allergic to penicillin). In clinical medicine, it seems, a non-reductive, dispositionalist conception of causation should be adopted.

The sufficient-cause model is useless to the clinician; however, both Rothman's conception and the Mill/Mackie account of causation from which it is derived, are very important to medicine. In the next chapter I show that, unlike the POA, the sufficient-cause model successfully accommodates nonmanipulable causes in epidemiology. In this chapter I showed that the full/sufficient cause of a disease, might well ground a useful method of disease-classification.

Notes

1   The two additional problems I raise have adapted from common objections in the metaphysics of causation literature (Mackie 1974; Lewis 1973a).
2   *Enquiry*, Sect. VII, Pt. II, 146.
3   These worlds are causally isolated from ours, whereby there is a (potentially infinite) plurality of universes.
4   Please note that this short paragraph misses out *many* of the subtleties of Lewis's view, and indeed misrepresents his view in some respects. I aim only to get the very basic gist of his approach to my reader. For a better understanding of Lewis's views, refer to his 1973a; 1973b; 1979; 2000. Daniel Nolan's (2005) provides an excellent summary.
5   My emphasis.
6   My emphasis.

DOI: 10.1057/9781137552921.0010

7   See also Collins et al (2004).

8   I discuss the methodology Kerry et al are referring to here in far greater depth in Chapter 4.

9   See Chapter 4.

10   How the question distinguishes between the conditions and the causes is unclear.

11   Risk is a statistical notion. The risk of an outcome is the number of people in the experimental group in which the outcome obtained/the number of people in the experimental group at the start of the study.

12   My emphasis.

13   It is perhaps worth noting (purely out of interest) that relatively few species of mosquito can transmit parasites of the genus *Plasmodium* (that which causes malaria in humans) – the most common genus of malaria carrying mosquito, *Anopheles*, includes more than 450 different species, only 30–40 of which can transmit malaria to humans.

14   Mumford and Anjum do not think dispositions and powers are interchangeable concepts, since according to them, dispositions are 'clusters of powers' (Mumford, 2004, 168–179; Mumford and Anjum, 2011, 99–103). However, in the philosophy of medicine literature, philosophers focus on dispositionality (Scheuermann et al 2009; Kerry et al 2012), and Mumford and Anjum's vector model is equally useful for modelling dispositions.

15   Whereby either the 'totality of all processes through which a given disease instance is realised'... (a) does not terminate in a return to normal [health] and (b) would, absent intervention, fall within [a disease] range... [or] (a) does not terminate in a return to normal [health] and (b) would, absent intervention, involve an increasing deviation from [health]. (Scheuermann et al, 2009, 118).

16   And later developed by Mackie.

17   With kind permission from Springer Science+Business Media: *Theoretical Medicine and Bioethics*, On the Classification of Diseases, 2014, August; 35(4), 251–269.

18   For now, consider a causal condition to be an event causally relevant to the contraction of the disease in question.

19   'Mask' is used here in Johnson's (1992) sense – exercise might mask a patient's disposition to get CVD just as dampening a match masks its disposition to light when struck.

20   This is integral to J.L. Mackie's (1974) conception of causation, where Mackie identifies a cause as any 'insufficient but non-redundant part of an unnecessary but sufficient condition' (an *inus* condition), that is, any one condition to be found within what he terms, the 'full cause'.

21   *System of Logic*, Book III, Ch. 5, Sect. 3.

22   There are also possible ways of amending the account to allow for *real* chance (that is, allowing that even upon knowing all states of affairs at *t* and all the laws of nature, it would be impossible to know for certain the state of affairs at *t*+1 due to objective probabilities in nature). One can give probability-raising accounts of causation, where the causal conditions are such that they are necessary to raise the probability of the effect in the circumstances. This would, in fact, be sufficient for my thesis. However, here I am happy to endorse metaphysical determinism and frequentist accounts of probability.

23   From personal correspondence with Alexander Broadbent, 2013.

24   See also Ellis and Lierse (1994).

DOI: 10.1057/9781137552921.0010

4

Population Health and Causal Inference

Abstract: *This chapter is concerned with the theoretical nature of causal inferences and effect measurement in epidemiology. Following Bird, I outline Hill's nine criteria for causal inference in medicine, and demonstrate why evidence-based medicine deems well-conducted experimental studies to be a more reliable means of identifying causation than observational studies. I then present two models of causal inference commonly employed by epidemiologists. First, the potential outcomes approach is put forward as a means of identifying practicable (Woodward 2004), manipulable causes of disease, and measuring the strength of the causal effects of exposures. We shall see that this approach does not accommodate all causal factors relevant to epidemiological practices. To resolve this issue, I suggest supplementing the potential outcomes approach with Rothman's (1976) sufficient-cause model.*

Keywords: Alex Broadbent; inferring causation; Nancy Cartwright; Philosophy of epidemiology; potential outcomes approach; the sufficient-cause model

Smart, Benjamin. *Concepts and Causes in the Philosophy of Disease.* Basingstoke: Palgrave Macmillan, 2016.
DOI: 10.1057/9781137552921.0011.

DOI: 10.1057/9781137552921.0011.

Identifying causal relationships: causal questions for epidemiologists

In the Introduction I defined epidemiology as the study of disease in populations, for the purposes of improving public health. This project is inextricably tied to the concept of causation, since in order to improve public health, one must (among other things, of course) identify the determinants of disease. To this end, epidemiologists have developed many types of experimental and observational studies, and a wide variety of statistical techniques. Correlations are relatively easy to discover, but discovering a correlation is not the same discovering a causal relationship. For example, drinking is correlated with lung cancer, but it does not cause lung cancer. It is correlated with lung cancer because drinkers are more likely to smoke, and smoking causes lung cancer. So a major part of the epidemiologist's job, is to determine which correlations are causal, and which are not. Furthermore, epidemiologists do not merely wish to establish causal relationships between exposures and outcomes, but how strong those causal relationships are; that is, how much of a difference the exposures makes to the outcomes.[1]

This chapter is concerned with the theoretical nature of causal inferences and effect measurement in epidemiology. Following Bird, in the next section I outline Hill's nine criteria for causal inference in medicine, demonstrating some of the problems epidemiologists face in making causal inferences. I then present two models of causal inference commonly employed by epidemiologists. First, the Potential outcomes approach (POA) is put forward as a means of identifying practicable (Woodward 2004) causes of disease and measuring the strength of the effects of exposures; I then question how successful this approach to causal inference in epidemiology really is. We shall see that the POA does not accommodate all causal factors relevant to epidemiological practices; to resolve this issue, I suggest supplementing the POA with Rothman's (1976) sufficient-cause model.

Hills criteria and the causal inference problem

Epidemiologists use a number of different kinds of study to identify causal relationships. The majority of these studies are 'observational'

DOI: 10.1057/9781137552921.0011

studies, such as cohort[2] or case-control studies,[3] but merely observing correlations is not enough to infer causation.

Addressing this issue, Hill famously listed nine criteria for distinguishing between causation and mere correlation or association (295–300); however, we shall see that these criteria are often insufficient to establish causation.

1 **Strength.** Hill provides two examples to illustrate that the stronger the association between exposure and outcome, the more justified one is in identifying a causal relationship: 'the mortality of chimney sweeps is 200 times that of workers who were not specially exposed to tar or mineral oils'; and 'the death rate from cancer of the lung in cigarette smokers is nine to ten times the rate in non-smokers'.

2 **Consistency.** This important criterion emphasises the need for causal associations to be repeatable, not only in the same location and at similar times, but under a wide variety of circumstances.

3 **Specificity.** The more specific/well-defined the exposures and outcomes, the more justified one is in inferring a causal relationship.

4 **Temporality.** 'Which is the cart and which is the horse?' In order to distinguish between cause and effect, one must consider which event precedes the other.

5 **Biological gradient.** If there is a causal relationship between the exposure and outcome, often there will be an observable biological gradient. In the case of cigarette smoking, for example, the more cigarettes one smokes, the more likely one is to get lung cancer.

6 **Plausibility.** The biological plausibility of a causal inference is determined by the accepted biological theories of the time. Of course, what is believed by scientists often changes over time, so consistency with the current popular theories in biology is not a necessary condition for inferring a causal relationship, but it is nonetheless desirable.

7 **Coherence.** Although biological plausibility cannot be demanded, an inferred causal relation 'should not conflict with the generally known facts of the natural history and biology of the disease'.

8 **Experiment.** Hill writes: 'Occasionally it is possible to appeal to experimental, or semi-experimental, evidence. For example, because of an observed association some preventative action is taken.' He asserts that this provides the strongest support for causal inference, and indeed, controlled trials like RCTs are typically deemed the most reliable means of establishing causation (Howick 2011).

DOI: 10.1057/9781137552921.0011.

9 **Analogy.** If two exposures, *xo* and *x1* are sufficiently similar, and there is a known causal relationship between *xo* and outcome O, it is sometimes reasonable to infer a causal relationship between *x1* and O.

Hill does not claim that any of these criteria (with the possible exception of specific kinds of well-designed and well-conducted experiment) individually succeed in identifying cause-effect relationships, but that each provides partial support for a causal link obtaining is fairly intuitive. Nonetheless, particularly in the case of observational studies, even close adherence to Hill's criteria will often result in spurious conclusions.

In an observational study, the epidemiologist records information about the subjects under study, taking measurements and asking questions, without actively intervening in their lives, or in their environments (Saracci, 2010, 10; Broadbent, 2013, 5–7). Here I show how Hill's criteria are applied to observational studies, but conclude that, even when these criteria are fulfilled, the observational studies so often employed by epidemiologists, are not always conclusive.

To establish a causal theory (T) – that there is a causal relationship between an exposure E and outcome O – one must eliminate: (R) the reverse hypothesis (that O causes E); (C) all competing hypothesis (e.g. common causes); and (N), the null hypothesis (Bird 2011). Bird argues that, in observational studies, Hill's criteria of strength, consistency, specificity, and biological gradient, support (T) by eliminating the null hypothesis. In other words – strong correlations between E and O; correlations between E and O in different circumstances; well-defined exposures and outcomes; and observable biological gradients – all increase the likelihood that the NHST will deem the correlation significant.

The temporality condition in an observational study supports T by eliminating the reverse hypothesis, since effects do not precede their causes.[4] However, the strategy outlined above is one of eliminative induction (Bird, 2011, 242), and therefore, in order to establish a clear causal relationship, one must also eliminate (C); that is, one must rule out every possible alternative explanation for the observed correlation between E and O. Consider the following observational study:

To test whether drinking caffeine causes bowel cancer, a case-control study is conducted in a hospital. Let us suppose the study reveals a higher risk of exposure to large quantities of caffeine among the bowel cancer group, than among the control group (drawn from the general population). Can one infer a causal relationship between drinking caffeine and bowel cancer? There are good reasons to think not, since *many* alternative

DOI: 10.1057/9781137552921.0011.

explanations cannot be ruled out. Those who drink excessive quantities of caffeine might work longer hours, be more stressed, work in hospitals, smoke more, drink more alcohol, eat more red meat, and so on – these are all *possible* explanations for the higher incidence rate. Furthermore, even if we seek to rule out every possible alternative explanation, how could we have done this? Perhaps there are common causes we haven't thought of. The application of Hill's criteria to an observational study, then, cannot eliminate C, since these criteria 'fail to exclude the common cause hypotheses in a *systematic* way' (Bird, 2011, 244).

Examples such as this highlight the fallibility of the epidemiologist's causal inferences, particularly in the case of observational studies.[5] In the next section, however, I investigate what has become a topical discussion in the philosophy of epidemiology (Broadbent 2015) – a theoretical means of inferring causal relationships in epidemiology, based on counterfactuals.

The epidemiologist's potential outcomes approach

In this section I present the epidemiologist's counterfactual, or potential outcomes approach (POA) to causation – this method of identifying causal relationships shares qualities with the counterfactual conceptions found in traditional analytic philosophy (Lewis 1973a), as the methodology presupposes that the effect of an exposure is measured relative to some contrary-to-fact condition.[6] In the next subsection, I first outline the general POA strategy, and how it relates to the counterfactual strategies most analytic philosophers are acquainted with; then I argue that although there are similarities, the links between the philosophical and epidemiological counterfactual approaches to identifying causal relationships are somewhat tenuous. I then demonstrate that for some common parameters (such as those presented in Cartwright's 2007 paper on RCTs), if the POA does succeed as a means of identifying relevant causal relationships, it does so using a frequentist probability raising account. I present Broadbent's characterisation of Hernan and Taubman's POA, and outline why he believes the POA involves both circular and false hypotheses – I then attempt to refute these objections.

I conclude that the POA strategy is not designed to identify all 'in principle' (Woodward 2004) causes of disease. It is a tool designed

DOI: 10.1057/9781137552921.0011.

to measure the effect of practicable interventions, and as such, does not deal with nonmanipulable causes (which are very often of fundamental importance to both epidemiological studies and clinical medicine). Given that the POA theoretical framework cannot accommodate nonmanipulable causes, I propose that in *some* circumstances, Rothman's sufficient-cause model is more suitable for the epidemiologist.

The potential outcomes approach methodology

The POA is concerned with many 'potential outcomes' (whereby outcomes can be a number of variables, including incidence rates, life expectancy, etc.), the values of which are determined through (a) actual group studies, and (b) estimates of the outcomes of counterfactual studies; that is, not only whether the actual outcome occurs given the non-occurrence of the actual exposure, but specific data concerning outcomes under a number of possible contrary-to-fact exposures. The strategy thus runs roughly as follows:

There are a number of possible actions, only one of which is actual. Take the actual action of an individual or population to be *xo*, and all alternative exposures to that individual or population to be uniquely specified: *x1, x2, x3*, and so on.

i   Take O to be a measure of outcome, and O*(xo)* to be the outcome of the observable event *xo*. O*(xn)* for all values of *n* except *o* are unobservable – they are the counterfactual outcomes (that is, what would have occurred were *x1, or x2*, or ... to have occurred). All outcomes O*(xn)* are potential outcomes (including the actual outcome).

ii Compare the actual outcome O(*xo*) with any one counterfactual outcome O(*xc*), to measure the value of *xo* versus O(*xc*).

Outcomes other than O*(xo)* are contrary-to-fact and thus unobservable, but reasonable estimation (assuming this is possible) of these values admits of numerous effect-measures (e.g. if one takes *xo* to be 'immunised against yellow fever', and *x1* to be 'not immunised against yellow fever', one can calculate the risk of yellow fever ratio due to non-immunisation versus immunisation by O*(x1)*/O*(xo)*).

The strategy employed implies that it is meaningless to assert that 'non-immunisation is a cause *simpliciter* of yellow-fever' – one must first determine which intervention/action it is relative to: one must

DOI: 10.1057/9781137552921.0011.

assert that 'non-immunisation causes yellow fever *versus immunisation*'. That one must always assert what an exposure is relative to, leads to some surprising consequences. For example, unprotected penile-vaginal sex (where HIV is prevalent) *is* a cause of HIV versus abstinence; but it is *not* a cause of HIV versus unprotected anal sex (where HIV is prevalent) – indeed, relative to unprotected anal sex, unprotected penile-vaginal sex reduces risk of HIV.

The potential outcomes approach and the Lewisian conception

Epidemiologists are usually concerned with studies confirming or falsifying *general* causal claims such as 'smoking causes cancer', or 'the *Plasmodium vivax* malaria transmission is more resilient to interruption than other forms of malaria' (Mendis et al 2009) – not so for the most influential philosophers espousing counterfactual conceptions of causation; they are more often interested in well-specified token events (e.g. '*Joe's* contracting malaria') – let us call this approach the 'Lewisian conception'.[7] Given that the epidemiologist is concerned with general causation, and the Lewisian with singular causation, the ties between Hume and Lewis's counterfactual conditional accounts and the epidemiologist's POA are tenuous in many respects. Nevertheless, proponents of the POA adopt a similar strategy to the Lewisian, in that they judge causal effects based not only on the actual outcomes of a patient's actual exposure, but also on the potential outcomes of alternative, unrealised exposures on the same patient(s) – assigning the phrase 'counterfactual' account thus seems appropriate to both the Lewisian conception, and the POA.

As we saw in Chapter 3, the Lewisian considers causation to be a matter of counterfactual dependence, whereby counterfactuals are evaluated using his 'possible worlds' ontology. This conception is of little use to the epidemiologist, of course, since moving from just a single causal inference (of the kind identified by the Lewisian method), to more general causal claims, is unjustified. Suppose, for example, that unbeknownst to Bob, Bob had always had a black mamba living under his bed. Bob habitually got out of bed *quietly* in the mornings, and consequently never woke the snake. Although it is true that getting out of bed quietly was a cause of Bob living a long and happy life, the general causal claim 'getting out of bed quietly increases life expectancy' is clearly false.

DOI: 10.1057/9781137552921.0011.

Reichenbach and the potential outcomes approach

Identifying causes in the way proposed by the POA (that is, via risk-parameter measurement (and comparison)), on the other hand, *can* be a notational variant of the frequentist probability-raising account of cause, similar to that proposed by Hans Reichenbach(1956).[8] To establish a causal relationship using risk-parameters, where x is a specified population, one measures whether an effect is more probable given a well-defined intervention by calculating the outcome $O(xo)$ (say, risk of morbidity), and the counterfactual outcome of the same parameter $O(x1)$, and comparing the two. Risk is equivalent to frequentist probability (the number of cases/population at the start of the study), so if the risk of morbidity given xo is higher than the risk given $x1$, that is, $P(D|xo)$ > $P(D|x1)$,[9] according to the Reichenbach probability-raising account of cause, xo is to be deemed a cause of disease D (given the nature of the POA, one must specify that xo is a cause relative to $x1$, of course). Note that, in the case of observational studies at least, this simplified model can present 'spurious correlations' as causal relationships, as a result of confounding factors (as highlighted by the 'alcohol causes lung cancer' example).

Proponents of the POA take this confounders problem to be adequately dealt with by randomisation (and where randomisation is not possible, similar methodological strategies) – the problem is thus a methodological one, to be dealt with through careful study-design. This short, one sentence response will hardly satisfy most readers, but the confounders problem is one that plagues epidemiological methods of all kinds.

Both Reichenbach and Lewis present their views as accounts of singular causation. They are clearly distinct analyses of causation, but (at least with respect to risk parameters) the POA is as similar to the Reichenbach view as it is to Lewis's account. The similarity to the Lewisian view rests on the use of counterfactuals, insofar as effect-measurements require statistics for outcomes under hypothetical scenarios. The similarity to the Reichenbach view rests on effect-measurements being made through the epidemiological equivalent to relative-probabilities; that is, by comparing the value of some parameter that plays the 'probability of the effect given the (proposed) cause'-role, and the value of the same parameter equivalent to the 'probability of the effect without the (proposed) cause'. Note, however, that not all parameters used by epidemiologists are equivalent to frequentist probability, so not all effect-measures made possible by

DOI: 10.1057/9781137552921.0011.

the POA will be notational variants of the probability-raising account. One can measure the effects of smoking in terms of life expectancy given smoking versus life expectancy given non-smoking, for example, but life expectancy is not a risk-measurement expressed in terms of frequentist probabilities.

I do not invite the reader to identify the numerous counterexamples to my simplistic exposition of Reichenbach's probability-raising conception, or that of Lewis's counterfactual theory (largely as more comprehensive expositions of both these are more defensible).[10] The expositions I have provided of both views, however, are sufficiently close to the more refined versions to establish the following claims: (i) that the Lewisian conception permits identifying cases of singular causation, whereas the epidemiologist must, just for pragmatic purposes, identify cases of general causation; (ii) the Lewisian conception and the POA are alike insofar as both involve consideration of contrary-to-fact suppositions, as well as one observable outcome; (iii) the POA only provides effect-measures *due* to the exposure *versus* an alternative specified (counterfactual) condition; and (iv) that in essence, when using risk-measurements, a POA analysis of cause can be viewed as a probability-raising account similar to the Reichenbach's, since risk in epidemiology is a notational variant of Reichenbach's frequentist probability; but again, the POA differs insofar as it is concerned with general causal statements, and there are alternative parameters epidemiologists use (such as life expectancy) which do not fit well with a probability-raising conception.

Hernan and Taubman's potential outcomes approach

Epidemiology is a prescriptive discipline, directed at improving public health through group studies. Given its prescriptive nature, prima facie epidemiologists need only be concerned with manipulable conditions – if one cannot prevent a casual condition being satisfied, and one cannot interfere with a casual condition as a form of treatment, one may as well consider it not a cause at all. To illustrate: it would be odd to consider 'being male' a cause of testicular cancer, even though only males contract the disease (in Mackie's eyes, being male would simply be a part of the 'causal field').[11] It might be sensible, however, to study the relationship between testosterone levels and testicular cancer, as testosterone levels can be interfered with. On the face of it, then, information concerning causal relationships between manipulable conditions (like testosterone levels) and diseases is, unlike information about nonmanipulable

DOI: 10.1057/9781137552921.0011.

conditions like 'being male', useful when deciding public health policy. It is this thought that drives Hernan and Taubman (2008) to assert that one cannot make claims about causal effects (within the epidemiological context) without specifying at least one well-defined intervention.

Here I criticise what I deem to be a naïve interpretation of the POA, by highlighting three objections raised in Alex Broadbent's (2015) paper on causation and prediction in epidemiology. Broadbent considers Hernan and Taubman's (two POA theorists) 2008 paper on obesity, in which they claim that no causal questions about the effects of obesity are meaningful. Broadbent takes this paper (along with significant additional evidence) to capture the essence of the POA, identifying four theses he deems advocates of the POA to be committed to: one semantic, one metaphysical, one pragmatic, and one epistemic.

Broadbent's characterisation of Hernan and Taubman's potential outcomes approach

The semantic thesis states that our pre-theoretic intuitions about causation should be ignored when developing an account of causation within epidemiology, acknowledging that it is the pragmatic consequences of the causal-conception that marks its quality. The semantic thesis thus captures the 'improving public health' prescriptive nature of the POA (so there is nothing obviously incoherent about it).

The metaphysical thesis states that 'at least sometimes, causes are difference-makers. X is a difference-maker for Y if and only if, had X had been absent or different, then Y would have been absent or different' (Broadbent, 2015, 76). This is a somewhat weaker version of Hume's conditional definition of cause, in which he state's 'if the first object had not been, the second never existed' (Hume 1999),[12] since causes do not (necessarily) always make a difference under the POA criterion – although smoking causes cancer, it might be the case that an individual would have developed cancer regardless of whether or not the individual smoked (perhaps due to some genetic predisposition). The weakening of the counterfactual thesis maintains the POA's commitments, without falling foul of obvious objections against anything stronger. This weakening is necessary due to the move from singular to general causation (but there is no need to explore this any further).

The pragmatic thesis states that 'the only difference-makers that epidemiology needs to care about are those that are humanly manipulable' (Broadbent, 2015, 76). This aspect is of course closely linked with the

DOI: 10.1057/9781137552921.0011.

semantic thesis, as prima facie for epidemiologists, usefulness is largely determined by what can *humanly* be changed. Unlike the semantic and metaphysical theses, Broadbent challenges the pragmatic thesis:

According to the pragmatic thesis, in order to know whether some condition is causal one must first know whether that condition can be (humanly) manipulated. This knowledge, says Broadbent, can only come from one of two sources. The first is background scientific knowledge, but this always involves causes without humanly manipulable interventions. The second is via direct empirical evidence: one attempts to manipulate some condition under investigation, and where one discovers that it does not correspond to a well-defined intervention, one declares it non-causal. Broadbent argues that in order to conduct such an experiment, one needs to have already identified a well-defined intervention, and since one does not know in advance whether an intervention is well-defined, the pragmatic thesis is flagrantly circular.

Finally, the epistemic thesis states that 'causal knowledge yields predictive knowledge.... *Enabling prediction under hypothetical scenarios is the mark of causal knowledge*' (Broadbent, 2015, 77). According to Broadbent this epistemic thesis can also be refuted, this time simply on the grounds that it is demonstrably false. It is undeniable that 'smoking causes cancer' is a useful causal claim, yet this knowledge has yielded many false predictions – for example, smoking low-tar cigarettes, or taking shallower puffs, is better for smokers than smoking standard cigarettes in the normal way.

In the following section I argue that although Broadbent's criticism of the Hernan and Taubman (and in particular the pragmatic thesis) is understandable (for reasons I demonstrate later in this chapter), the thesis epidemiologists are predominantly interested in, the semantic thesis, warrants rethinking the pragmatic and epistemic theses such that they avoid the circularity and falsity objections. In the following sections I suggest that applying the Popperian scientific method resolves Broadbent's worries; furthermore, I show that causal knowledge obtained via the POA does in fact yield at least some predictive knowledge.

Diffusing Broadbent – A Popperian take on the potential outcomes approach

Broadbent's objection to the pragmatic thesis is predicated on Hernan's inability to answer 'how can the epidemiologist know which conditions

DOI: 10.1057/9781137552921.0011.

are humanly manipulable?' without succumbing to a vicious circle. As we saw, the two possible approaches to answering the question (as identified by Broadbent) are (i) existing scientific background knowledge – but this fails because it involves 'a large quantity of causes that do not correspond to humanly manipulable interventions' (77), and (ii) empirical investigation – but this fails as 'for good empirical enquiry, you need *already* to have identified a well-defined intervention' (77). However, although Broadbent is right to claim only existing scientific knowledge and empirical enquiry can provide the information, this is not problematic.

The paper Broadbent focuses on questions whether obesity shortens life (Hernan and Taubman 2008). Hernan and Taubman conclude that obesity cannot be considered a cause because it does not correspond to a well-defined intervention – demonstrable, in this case, because there are many ways to lose weight,[13] each of which has a distinct effect on mortality; that is, empirical studies designed to measure death attributable to obesity give different values depending on the mode of intervention. Given that there is no well-defined intervention, obesity is not a condition suitable for causal effect measurement.

Although the Hernan/Taubman paper is prima facie convincing regarding the non-causal status of obesity, Broadbent points out that 'if we [discover whether something corresponds to a well-specified[14] intervention] by empirical investigation, then we cannot have decided *in advance*[15] whether an intervention is well-specified ... Thus we cannot make the question of whether our empirical investigation is a good one depend on advance knowledge of whether the putative causes we are contemplating correspond to well-specified interventions' (2015, 78).

Although Broadbent's argument is sound, such a strategy is only problematic under certain assumptions – the key question being: 'to what extent must we *know* that an intervention is well-defined prior to investigation?' Suppose that instead of reading the POA to assume causes require 'confirmed' well-defined interventions, one is more charitable, and takes Hernan to assume epidemiologists adopt a broadly Popperian strategy; that is, rather than taking an hypothesis to be unequivocally proven, they use our best scientific theories, and analogies with similar cases of better-corroborated well-defined interventions, to determine (with a well-educated guess) which potential causes have well-defined interventions. Of course, Broadbent is right in his assertion that observation must come subsequent to theory, but this passage from

DOI: 10.1057/9781137552921.0011.

Popper's 1972 suggests that this fact might not be problematic, but quite usual in the sciences.

> The belief that science proceeds from observation to theory is still so widely and so firmly held that my denial of it is often met with incredulity... But in fact the belief that we can start with pure observation alone, without anything in the nature of theory, is absurd... Observation is always selective. It needs a chosen object, a definite task, an interest, a point of view, a problem. And its description presupposes a descriptive language, with property words; it presupposes similarity and classification, which in turn presupposes interests, points of view, and problems. (Popper, 1972, 46)

On a Popperian, falsificationist view, the epidemiologist can assert that an empirical investigation is a good one if our best scientific theories identify the relevant intervention as 'well-defined' in advance of the empirical investigation, while maintaining an appropriate level of scepticism (and attempting to falsify the hypothesis through empirical investigation).

Suppose, for example, we did not know that smoking could lead to weight loss, and our best guess is that the only way of losing weight is to eat healthily. Although we could not know for sure that obesity (the proposed cause of, say, heart disease) corresponds to a well-defined intervention (which, it later turns out, it doesn't), that is not to say we were not justified in treating it as if it does, given our 'knowledge' at the time (later, this hypothesis would be falsified, but that's OK). One is not asserting that obesity definitely corresponds to a well-defined intervention, only that, given our current understanding of similar cases, it seems likely that it does – there is no circularity here. Of course, once science reveals that obesity does *not* have a well-defined intervention, it is no longer a candidate for a cause of disease, but that does not mean the initial investigation was a bad one. A broadly Popperian attitude to scientific investigation, then, diffuses Broadbent's circularity objection to the POA.

What of the falsity objection to the epistemic thesis: that '*Enabling prediction under hypothetical scenarios* is the mark of causal knowledge' (6), yet the POA conception of causation regularly yields false predictions? I do not think one should be concerned by this objection, either. On the face of it, the objection is only valid if 'enabling prediction under hypothetical scenarios' *requires* the predictions to come true – of course, perfectly justified predictions can (and often do) turn out false, so this cannot be what Broadbent has in mind. Indeed, the objection is clarified

DOI: 10.1057/9781137552921.0011

in a footnote. Referring to the false predictions made about what happens when tar in cigarettes is lowered, he states:

> Strictly speaking... I should say that these predictions were not only false but also bad in some more substantive way. In particular... the implementation of causal knowledge in making those predictions did not constitute checking how one might be wrong, and thus did not by itself amount to a good prediction activity (2015, 78)

Although Broadbent is right to highlight the importance of the epistemic thesis in the POA, his refutation relies on his own conditions for what it is to be a good prediction activity. Clearly, true predictions can fall out of a poor prediction activities (wild guesses can turn out to be true), and false predictions can fall out of good prediction activities. However, if Broadbent's objection is to hold any weight, he would have to deny the following statement: 'If, following a population of fast food eaters, one knows via the POA that the counterfactual outcome O(x healthy diet) is a high life expectancy versus actual outcome O(x fast food diet), then one can predict (well) that if a population changes their eating habits from a fast food diet to eating healthily, their life expectancy increases.' The causal statement alone does not consider alternative outcomes to the counterfactual scenario, and thus according to Broadbent's criteria, the prediction activity is a bad one – a counterintuitive conclusion indeed (in fact, the relation between the causal and predictive claims looks to be a matter of logical entailment). Let us consider Broadbent's example to see why this comes about:

> As a descriptive claim, the epistemic thesis is clearly false, as there are plenty of cases where we have what we might call 'causal knowledge' but make bad predictions about what will happen under hypothetical suppositions. Smoking causes lung cancer, yet predictions based on the knowledge about what would happen if the amount of tar in cigarettes were lowered, or if smokers took shallower puffs were mistaken. (78)

Take A to be 'smoking with normal puffs'; B to be 'lung cancer'; and C to be 'smoking with shallow puffs'. The structure of the inference above thus runs as follows: we know that 'A raises the probability of B' (Smoking with normal puffs causes lung cancer), therefore we can justifiably infer 'C raises the probability of not-B' (if I take shallower puffs then I am less likely to get lung cancer). This is quite clearly a fallacious move, since A and C are distinct activities (despite both involving smoking, C is not the negation of A). No sensible advocate of 'causal knowledge yields

DOI: 10.1057/9781137552921.0011.

predictive knowledge' would think this inference infallible. Good prediction activity, when seen from a causal perspective, requires the *right* causal knowledge. We know that smoking causes cancer versus non-smoking, and, as required by the epistemic thesis, this yields predictive knowledge. Specifically, it enables predictions such as 'if this population stops smoking, they are less likely to get lung cancer,'[16] but there is a limited amount of predictive knowledge that any one causal claim can yield. The predictive knowledge is, in fact, sometimes determinately specified by the counterfactual scenarios considered in the POA study: If smoking causes cancer versus not smoking, then one can predict stopping smoking will reduce risk of lung cancer. If taking ibuprofen reduces headache versus taking placebos, then one can predict taking the ibuprofen over the placebo will reduce the headache. The move from 'ibuprofen reduces headache versus placebo' to 'my headache will go away faster if I take ibuprofen, than if I take paracetamol,' however, is a bad prediction activity, as the POA causal claim (remember that POA causal claims are meaningless without specifying a contrast) makes no reference to any counterfactual scenario involving paracetamol. The POA does enable prediction (albeit without guaranteeing the prediction will be true), and given the very nature of POA causal claims, POA causal knowledge *always* yields *some* predictive knowledge. Of course, the strategy the POA employs makes it possible for one to be incorrect about effect-measures, as estimations of outcomes under counterfactual scenarios can be very wrong (one might, for example, be completely unaware of some condition that would affect the outcome of the counterfactual scenario), but this is a problem with the POA's overall strategy, rather than the epistemic thesis. Hernan and Taubman should not, then, be overly worried by Broadbent's charge.

The importance of nonmanipulable causes

Nancy Cartwright, in her 2010 PSA presidential address criticising RCTs, demonstrated that in practice, unjustified extrapolations from 'intervention X works somewhere' to the claim that the 'intervention X works in general' are regularly made. She highlights two instances of nutritional intervention funded by The World Bank. The first was conducted in Tamil Nadu state, India, and comprised providing food supplements, health measures, and educating pregnant mothers about how to better nourish their children. The project was very successful, with malnutrition dropping significantly in the region. The same programme was then implemented in Bangladesh – an area with similar problems. Given the

DOI: 10.1057/9781137552921.0011.

success in Tamil Nadu, a POA study would suggest a similar drop in malnourished children elsewhere, but in Bangladesh there was no such drop. So what explains the failure of the inferred general causal claim? Two reasons were proposed: first, 'food supplied by the project was used not as a supplement but as a substitute, with the usual food allocation for that child passing to another member of the family' (Cartwright 2010, 23), and second, that 'the program targeted the mothers of young children. But [in Bangladesh] mothers are frequently not the decision makers... with respect to the health and nutrition of their children' (White, 2009, 6, in Cartwright, 2010, 24).

In Bangladesh, the background conditions were such that the intervention funded by The World Bank was insufficient to reduce malnutrition, since even with the extra food and education, conditions together sufficient for malnourishment were in place. On the other hand, in Tamil Nadu, educating young mothers removed a necessary condition of child-malnourishment, so the program was successful. This study implies that in certain circumstances, an alternative to the POA causal model, like the conception provided by Rothman below, must be applied.

As one can see through the example above, an effect is often not determined by a single causal factor – indeed, usually there are several states of affairs that are only together sufficient for the effect. This is captured by the sufficient-cause model (Rothman 1976) outlined in the previous chapter, according to which a sufficient cause is represented by a pie-chart.[17]

Figure 4.1 is the pie chart for child-nutrition – it shows that the sufficient cause for a well-nourished child includes sufficient food, a well-educated nutrition decision-maker, the correct distribution of the food, and 'other' unspecified non-redundant conditions. Each segment of the chart is a necessary condition for good child-nutrition. Prior to the intervention, in both Tamil Nadu and Bangladesh segments of the pie were missing. In Tamil Nadu there was insufficient food and the decision maker (the mother) was insufficiently educated in child-nutrition. The project funded by The World Bank implemented interventions such that the sufficient cause was completed and the child-malnutrition levels dropped. In Bangladesh, however, educating the mothers in child-nutrition and providing food supplements did not have the desired effect, as only one of the absent conditions in the sufficient cause was satisfied. First, because educating the mother did not help with the 'educated

DOI: 10.1057/9781137552921.0011.

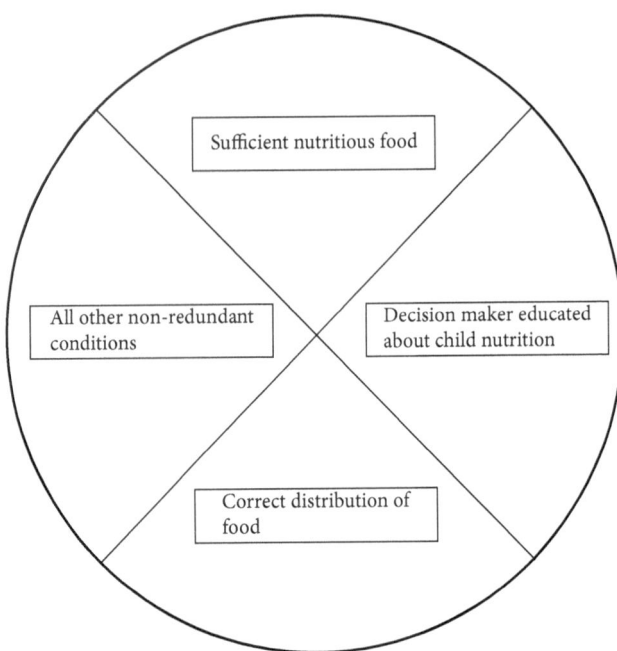

FIGURE 4.1   *Malnutrition's 'sufficient-cause'*

nutritional decision-maker' condition, and second because although there the project provided sufficient food, the household food was not being distributed in the manner intended.

The most important lesson here is that a sufficient-cause is unlikely to comprise only humanly manipulable states of affairs – it is highly unlikely, for example that an epidemiologist can, in practice, change the decision-making processes of an entire culture. Even if this were possible, among the 'other non-redundant conditions' will be factors that are nonmanipulable: the price of healthy food, whether or not there were floods or droughts, and so on. The epidemiologist must, when making decisions about studies to conduct (and what public health policies to suggest), take at least some nonmanipulable causes into consideration.

Nonmanipulable causes, epidemiology, and clinical decision making

Cartwright's example highlights the fact that outcomes and interventions are not the only important factors in assessing the kind of counterfactual

DOI: 10.1057/9781137552921.0011.

epidemiologists are interested in, as additional conditions shape the results of studies. In the obesity example, Hernan and Taubman argue that one should not consider obesity a cause of increased morbidity since obesity does not have a well-defined intervention – ultimately, this is because ill-defined interventions make assessing the counterfactual 'if she were not obese, she would have lived longer' impossible, since different interventions produce different truth-values. But the same is true of any intervention in which nonmanipulable factors affect the truth-value of counterfactuals, when these are not well-specified. For example, two people may have the same deadly bacterial infection, but one cannot necessarily state that 'if the patient ingests penicillin her risk of death is lower' when important, nonmanipulable conditions such as penicillin allergies have been ignored. Not only do the interventions need to be well-specified, but nonmanipulable background conditions, too – they 'make a difference'. The POA may be a thesis about interventions; a thesis driven by what epidemiologists and medical practitioners can actually *do* to prevent or treat diseases, but there is no questioning the causal relevance of nonmanipulable factors in both epidemiology and clinical medicine.

The POA is not a suitable candidate for identifying all causes of disease, even within an epidemiological framework. The structure of the approach provides effect-measures of manipulable conditions, and of course the effects being measured have causes – namely, the well-defined interventions $0, 1, 2 \ldots n$, but manipulable conditions are not the only relevant causes in epidemiology. Of course, that the POA provides a means of measuring the effect of one intervention versus another, and does not accommodate nonmanipulable states of affairs, does not rule nonmanipulable states of affairs non-causal. All one can conclude from the method outlined in this chapter is that the POA cannot measure the effect of nonmanipulable causes.

This conclusion does not suggest that *all* nonmanipulable but non-redundant states of affairs should be considered causes by those involved in medical research. Epidemiology is a prescriptive discipline, and thus only those events/states of affairs that it is *useful* to call causes should be considered causes. Nonetheless, there are many cases in which it is necessary to take nonmanipulable causes seriously, such as the case of malnutrition in Tamil Nadu and Bangladesh. There I used Rothman's sufficient-cause model to map the relevant causal factors – a suitable candidate, I think. But this would not be a suitable causal-model for

DOI: 10.1057/9781137552921.0011

measuring the causal effect of smoking on life expectancy. How one can best model causation in epidemiology, then, looks to change depending on context.

Notes

1   One of the epidemiologist's roles is to advise how best to allocate funds – in principle, one allocates the most money to dealing with exposures with very strong causal connections to common and devastating diseases. For example, epidemiologists discovered a very strong causal connection between smoking and lung cancer, and a relatively very weak (but nonetheless genuine) causal connection between air pollution and lung cancer. Unsurprisingly, more money is spent trying to get people to stop smoking, than trying to get people to avoid walking along busy roads.

2   In a cohort study, two groups are chosen at the beginning of the study, one with the exposure (e.g. smoking, a kind of medication, etc.) and the other without. After a specified period, the epidemiologist investigates the proportion of those groups with the outcome (e.g. cancer, health, a side effect, etc.), and the results are compared.

3   In case-control studies, a group with 'the outcome' (e.g. cancer) is studied, to determine in what proportion of the group the exposure obtained ('the risk of exposure' – e.g. what proportion of the cancer patients smoked). This result is compared with the 'risk of exposure' of a group of people in which the outcome does not obtain (e.g. a group of people not suffering from cancer).

4   I am sure some metaphysicians will question this (anyone who believes time travel is possible, for example), but in the context of medical research, this is a fair assumption!

5   Howick (2011) and Bird (2011) argue that experimental studies, like RCTs, are far less vulnerable to inferring spurious causal relationships.

6   When asking whether A caused B, Lewis asks us to consider a world in which everything is the same, aside from the fact that A does not occur (contrary-to-fact) – A causes B if, in this world, B does not occur.

7   Hume's work is often seen as the origin of counterfactual conditional accounts of causation – it may seem surprising, then, that I refer to the view as 'Lewisian'; I call the philosophical (as opposed to epidemiological) counterfactual analysis of causation the 'Lewisian counterfactual conception' for two reasons: first, because Lewis's account is (unsurprisingly, given a couple of centuries' progress in philosophy) more sophisticated and less open to counterexamples than Hume's, but second, and more importantly, because Hume provides a regularity approach to causation (for which,

DOI: 10.1057/9781137552921.0011.

it is probably fair to say, he is more often associated with by analytic philosophers) in addition to his counterfactual approach, not at all grounded by counterfactuals (if anything, counterfactuals are grounded by regularities). More on regularity approaches to causation in later sections.

8   RCTs certainly employ this kind of approach to causation.

9   To be read: The probability/risk of disease given xo is greater than the probability/risk of disease given xo.

10 If the reader is interested in these philosophical issues she should consult Reichenbach's (1956) and Lewis's (1973a and 2000) work directly. More detailed expositions of the POA can be found in Rothman, Greenland, and Lash (2008, chapter 4), and Greenland (2005)

11 See Chapter 3.

12 *Enquiry*, Sect. VII, Pt. II, 146.

13 Different diets, different exercise routines, smoking, and so on.

14 'Well-specified' and 'well-defined' should be taken as synonyms.

15 My emphasis.

16 This requires the accuracy of our counterfactual suppositions in the POA process of identifying causes, of course, but one must assume this.

17 See Chapter 3.

DOI: 10.1057/9781137552921.0011.

Conclusion: Context Dependency is Rife

Smart, Benjamin. *Concepts and Causes in the Philosophy of Disease*. Basingstoke: Palgrave Macmillan, 2016.
DOI: 10.1057/9781137552921.0012.

▶

DOI: 10.1057/9781137552921.0012

In this book I have attempted to provide analyses of the concepts and methods used by the medical profession. Medicine is ultimately a goal-directed discipline: to improve the health of the general public. This is true whether one is an epidemiologist discovering causes and effective interventions, a pathologist examining tissues for diagnostic purposes, or a clinician attempting to improve the wellbeing of the patients.

There are many different concepts of disease, and many different concepts of causation. Some of these are entirely unsuccessful, but some are successful in some contexts but not in others. A value-free concept of disease is useless for deciding whether or not an individual should be legally responsible for her actions on health grounds, since this requires an assessment of her state of wellbeing – an inherently value-laden notion. The pathologist does not make value judgements, however. The pathologist examines organs, tissues, and cells to determine whether they are performing their natural function to the required standard, and if not, why not – she does not need to consider whether the patient deems whatever condition he has to be a bad thing. In medicine, the 'right' concept of disease is that which is most useful at the time – it is entirely context dependent.

The same is true of concepts of causation. Notions of causation are of course fundamental to epidemiology. Some kinds of trial presuppose a very particular concept of causation. Cartwright states that RCTs are seen as the gold standard because they *imply* their causal conclusions – but this can only be asserted if one presupposes a probability-raising account of causation. The POA is great for measuring the effects of manipulable interventions, but the epidemiologist cannot forget that manipulable interventions are not the only type of cause. She must employ a completely different concept of causation, such as the sufficient-cause model, if she is to accommodate situations in which nonmanipulable factors play a role – that model, however, is no good for measuring causal effects.

Thus, the concept of disease that is useful to the clinician is different to that pertaining to pathology, and both are different again from that which is appropriate to the epidemiologist.

DOI: 10.1057/9781137552921.0012

Bibliography

Ananth, M. 2008. *In Defense of an Evolutionary Concept of Health: Natures, Norms and Human Biology*. Aldershot: Ashgate Publishing Limited.

Armstrong, D. 1983, What Is a Law of Nature?, Cambridge: Cambridge University Press.

Bird, A. 2007. *Nature's Metaphysics: Laws and Properties*. Oxford: Clarendon Press.

——. 2011. The Epistemological Function of Hill's Criteria. *Preventive Medicine*, 53(4–5): 242–5.

Boorse, C. 1975. On the Distinction between Disease and Illness. *Philosophy and Public Affairs*, 5(1): 49–68.

——. 1976. What a Theory of Mental Health Should Be. *Journal for the Theory of Social Behavior*, 6(1): 61–84.

——. 1977. Health as a Theoretical Concept. *Philosophy of Science*, 44(4): 542–73.

——. 1997. A Rebuttal on Health. In J. M. Humber & R. F. Almeder (eds.), *What Is Disease?* Totowa: Humana Press pp. 1–134.

——. 2002. A Rebuttal on Functions. In A. Ariew, R. Cummins, & M. Perlman (eds.), *Functions*, 63–112. New York: Oxford.

——. 2014. A Second Rebuttal on Health. *Journal of Medicine and Philosophy*, 39(6): 683–724.

Broadbent, A. 2008. The Difference between Cause and Condition. *Proceedings of the Aristotelian Society*, 108: 355–64.

——. 2009. Causation and Models of Disease in Epidemiology. *Studies in History and Philosophy of Biological and Biomedical Sciences*, 40(4): 302–11.

DOI: 10.1057/9781137552921.0013

———. 2012. Causes of Causes. *Philosophical Studies*, 158(3): 457–76.

———. 2013. *The Philosophy of Epidemiology*. London: Palgrave Macmillan.

———. 2015. Causation and Prediction in Epidemiology: A Guide to the Methodological Revolution. *Studies in History and Philosophy of Biological and Biomedical Sciences*, 54, 72–80.

Cartwright, N. 2007. Are RCTs the Gold Standard? *BioSocieties*, 2(1): 11–20.

Cartwright, N. *Will This Policy Work for You? Predicting Effectiveness Better: How Philosophy Helps* Transcript of 2010 The Philosophy of Science Association presidential address. http://sdcc3.ucsd.edu/~amarcell/index.php_files/cartwright_3.pdf

———. 2012. Will This Policy Work for You? Predicting Effectiveness Better: How Philosophy Helps. *Philosophy of Science*, 79(5): 973–89.

Collins, J., Hall, N., & Paul, L.A. 2004. *Causation and Counterfactuals*. Cambridge, MA: MIT Press.

Cooper, R. 2002. Disease. *Studies in History and Philosophy of Science*, 33(2): 263–82.

Dawkins, R. 1976. *The Selfish Gene*. Oxford: Oxford University Press.

DeVito, S. 2000. On the Value-neutrality of the Concepts of Health and Disease: Unto the Breach Again. *Journal of Medicine and Philosophy*, 25(5): 539–67.

Eke, P.I., Dye, B.A., Wei, L., Thornton-Evans, G.O. & Genco, R.J. 2012. Prevalence of Periodontitis in Adults in the United of States: 2009 and 2010. *Journal of Dental Research*, 91(10): 914–20.

Ellis, B. 2001. *Scientific Essentialism*. Cambridge: Cambridge University Press.

———. 2009. *The Metaphysics of Scientific Realism*. Durham: Acumen Publishing.

Ellis, B. & Lierse, C. 1994. Dispositional Essentialism. *Australasian Journal of Philosophy*, 71(1): 27–45.

Engelhardt, H.T. Jr. 1976. Ideology and Etiology. *Journal of Medicine and Philosophy*, 1(3): 256–68.

Ereshefsky, M. 2009. Defining 'Health' and 'Disease'. *Studies in History and Philosophy of Biological and Biomedical Sciences*, 40(3): 221–7.

Eriksen, T et al. 2013. At the Borders of Medical Reasoning: Aetiological Challenges of Medically Unexplained Symptoms. *Philosophy Ethics and Humanities in Medicine*, 8: 11.

Fisher, R.A. 1925. *Statistical Methods for Research Workers*. Edinburgh: Oliver and Boyd.

Flew, A. 1973. *Crime or Disease?* London: Macmillan Press.

DOI: 10.1057/9781137552921.0013

Geach, P. 1972. *Logic Matters*. Berkeley: University of California Press.

Giroux, E. 2015. Epidemiology and the Bio-statistical Theory of Disease: A Challenging Perspective. *Theoretical Medicine and Bioethics*, 36(3): 175–95.

Greenland, S. 2005. Epidemiologic measures and policy formulation: lessons from potential outcomes. *Emerging Themes in Epidemiology*, 2(5).

Guerrero, D.J. 2010. On a Naturalist Theory of Health: A Critique. *Studies in History and Philosophy of Biological and Biomedical Sciences*, 41(3): 272–8.

Hall, N. 2000. Causation and the Price of Transitivity. *Journal of Philosophy*, 97(4): 198–222.

——. 2004. Two Concepts of Causation. In Collins, N., Hall, N., & Paul, L.A (eds.), *Causation and Counterfactuals*, 225–76. Cambridge, MA: MIT Press.

Hausman, D.M. 1996. Causation and Counterfactual Dependence Reconsidered. *Nous*, 30(1): 55–74.

——. 2012. Health, Wellbeing, and Measuring the Burden of Disease. *Population Health Metrics*. 10(13): 1–7.

Heil, J. 2012. *The Universe as We Find It*. Oxford: Oxford University Press.

Hek, K. et al. 2013. A Genome-wide Association Study of Depressive Symptoms. *Biological Psychiatry*, 73(7): 667–78.

Hernan, M.A. 2005. Invited Commentary: Hypothetical Interventions to Define Causal Effects – Afterthought of Prerequisite? *American Journal of Epidemiology*, 162(7): 618–20.

Hernan, M.A, & Taubman, S.L. 2008. Does Obesity Shorten Life? The Importance of Well-Defined Interventions to Answer Causal Questions. *International Journal of Obesity*, 32. S8–S14.

Hesslow, G. 1993. Do We Need a Concept of Disease? *Theoretical Medicine*, 14(1): 1–14.

Hill, A.B. 1965. The Environment and Disease: Association or Causation? *Proceedings of the Royal Society of Medicine*, 58: 295–300.

Howick, J.H. 2011. *The Philosophy of Evidence-Based Medicine*. Hoboken, NJ: Wiley- Blackwell.

Hume, D. 1999. *An Enquiry Concerning Human Understanding*. Oxford: Oxford University Press.

Johnson, M. 1992. How to Speak of the Colors. *Philosophical Studies*, 68(3): 221–63.

DOI: 10.1057/9781137552921.0013

Kerry, R. et al. 2012. Causation and Evidence-based Practice: An Ontological Review. *Journal of Evaluation in Clinical Practice*, 18(5): 1006–12.

Kingma, E. 2007. What Is It to Be Healthy? *Analysis*, 67(294): 128–33.

———. 2010. Paracetamol, Poison and Polio: Why Boorse's Account of Function Fails to Distinguish Health and Disease. *British Journal for the Philosophy of Science*, 61(2): 241–64.

———. 2012. Health and Disease: Social Constructivism as a Combination of Naturalism and Normativism. In Carel, H., & Cooper, R.V. *Health, Illness and Disease: Philosophical Essays*, 37–56. Durham: Acumen Publishers.

———. 2014. Naturalism about Health and Disease: Adding Nuance for Progress. *Journal of Medicine and Philosophy*, 39(6): 590–608.

Kirk, R.E. 1996. Practical Significance: A Concept Whose Time Has Come. *Educational and Psychological Measurement*, 56: 746–59.

Kuo, H.K. & Fujise, K. 2011. Human Papillomavirus and Cardiovascular Disease among U.S. Women in the National Health and Nutrition Examination Survey, 2003 to 2006. *Journal of the American College of Cardiology*, 58(19): 2001–6.

Lewis, D. 1973a. *Counterfactuals*. Oxford: Blackwell Publishers.

———. 1973b. Causation. *Journal of Philosophy*, 70(17): 556–67.

———. 1979. Counterfactual Dependence and Time's Arrow. *Nous*, 13(4): 455–76.

———. 2000. Causation and Influence. *Journal of Philosophy*, 97(4): 182–97.

Mackie, J.L. 1974. *The Cement of the Universe*. Oxford: Oxford University Press.

Margolis, J. 1976. The Concept of Disease. *Journal of Medicine and Philosophy*, 1(3): 238–55.

Maslen, C. 2004. Causes, Contrasts and the Nontransitivity of Causation. In Collins, N., Hall, N., & Paul, L.A (eds.), *Causation and Counterfactuals*, 341–57. Cambridge, MA: MIT Press.

Matsushita, M., Uchida, K., & Okazaki, K. 2007. Role of the Appendix in the Pathogenesis of Ulcerative Colitis. *Imflammopharmacology*, 15(15): 154–57.

Mendis, K., Rietveld, A., Warsame, M., Bosman, A., Greenwood, B. and Wernsdorfer, W. H. 2009. From Malaria Control to Eradication: The WHO Perspective. *Tropical Medicine & International Health*, 14: 802–809. doi: 10.1111/j.1365-3156.2009.02287

DOI: 10.1057/9781137552921.0013

Mezies, P. 2004. Difference-Making in Context, in Collins, Hall, and Paul 2004, pp. 139–80.

Mill, J.S. 1941. *A System of Logic*. 8th edn. (reprinted), London.

Millikan, R.G. 1984. *Language, Thought, and Other Biological Categories: New Foundations for Realism*. Cambridge, MA: MIT Press.

——. 1989. In Defence of Proper Functions. *Philosophy of Science*, 56(2): 288–303.

Mumford, S. & Anjum, R. 2011. *Getting Causes from Powers*. New York: Oxford University Press.

Mumford, S., 2004, Laws in Nature, London: Routledge.

Neander, K. 1991. The Teleological Notion of 'Function'. *Australasian Journal of Philosophy*, 69(4): 454–68.

Putnam, H. 1975. *Mind, Language and Reality: Philosophical Papers Vol. 2*. Cambridge: Cambridge University Press.

Popper, K. 1972. *Conjectures and Refutations: The Growth of Scientific Knowledge*. 4th edn. London: Routledge and Kegan Paul.

Reichenbach, H. 1956. *The Direction of Time*. Berkeley and Los Angeles: University of California Press.

Rothman, K.J. 1976. Causes. *American Journal of Epidemiology*, 104: 587–92.

Rothman, K.J., Greenland, S., & Lash, T.L. 2008. *Modern Epidemiology*. 3rd edn. Lippincott: Williams & Wilkins.

Ruse, M. 1973. *The Philosophy of Biology*. London: Hutchinson University Press.

Ruse, M. 1987. Biological Species: Natural Kinds, Individuals, or What? *British Journal for the Philosophy of Science*, 38(2): 225–42.

Saracci, R. 2010. *Epidemiology: A Very Short Introduction*. Oxford: Oxford University Press.

Schaffer, J. 2005. Contrastive Causation, Philosophical Review, 114: 297–328.

Schaffer, J. 2012. Causal Contextualism. In Blaauw (ed.), *Contrastivism in Philosophy: New Perspectives*, 35–63. New York: Routledge.

Scheuermann, R.H., Ceusters, W., & Smith, B. 2009. Toward an Ontological Treatment of Disease and Diagnosis. *Summit on Translat Bioforma*, 116–20. http://www.ncbi.nlm.nih.gov/pmc/articles/PMC3041577/

Schwartz, P.H. 2007. Natural Selection, Design, and Drawing a Line. *Philosophy of Science*, 74(3): 364–85.

DOI: 10.1057/9781137552921.0013

Smart, B. 2014. On the Classification of Diseases. *Theoretical Medicine and Bioethics*, 35(4): 251–69.

Smith, J.M. 1998. *Evolutionary Genetics*. 2nd edn. Oxford: Oxford University Press.

Wakefield, J. C. 1999. Evolutionary versus prototype analyses of the concept of disorder. *Journal of Abnormal Psychology*, 108, 374–399.

Wakefield, J. 1992. The Concept of Mental Disorder: On the Boundary between Biological Facts and Social Values. *American Psychologist*, 47: 373–88.

Whitbeck, C. 1977. Causation in Medicine: The Disease Entity Model. *Philosophy of Science*, 44(4): 619–37.

White, H. 2009. "Theory-Based Impact Evaluation: Principles and Practice." 3ie Working Paper 3. New Delhi: International Initiative for Impact Evaluation.

Woodward, J. 2003. Making Things Happen: A Theory of Causal Explanation. Oxford: Oxford University Press.

Woodward, J. 2004. Counterfactuals and Causal Explanation. *International Studies in Philosophy of Science*. 18(1): 41–72.

Wright, L. 1976. Teleological Explanations. Berkeley, CA: University of California Press.

DOI: 10.1057/9781137552921.0013

Index

DOI: 10.1057/9781137552921.0014

DOI: 10.1057/9781137552921.0014

DOI: 10.1057/9781137552921.0014